Quantum Physics for Beginners

QUANTUM PHYSICS FOR BEGINNERS

The Step-by-Step Guide to Discover All the Mind-Blowing Secrets of Quantum Mechanics and How You Unknowingly Use Its Most Famous Theories Every Day

By Michael Rutherford

Quantum Physics for Beginners

© **Copyright 2022 - All rights reserved.**

The content contained within this book may not be reproduced, duplicated or transmitted without direct written permission from the author or the publisher.

Under no circumstances will any blame or legal responsibility be held against the publisher, or author, for any damages, reparation, or monetary loss due to the information contained within this book. Either directly or indirectly.

Legal Notice:

This book is copyright protected. This book is only for personal use. You cannot amend, distribute, sell, use, quote or paraphrase any part, or the content within this book, without the consent of the author or publisher.

Disclaimer Notice:

Please note the information contained within this document is for educational and entertainment purposes only. All effort has been executed to present accurate, up to date, and reliable, complete information. No warranties of any kind are declared or implied. Readers acknowledge that the author is not engaging in the rendering of legal, financial, medical or professional advice. The content within this book has been derived from various sources. Please consult a licensed professional before attempting any techniques outlined in this book.

By reading this document, the reader agrees that under no circumstances is the author responsible for any losses, direct or indirect, which are incurred as a result of the use of information contained within this document, including, but not limited to, errors, omissions, or inaccuracies.

Quantum Physics for Beginners

TABLE OF CONTENTS

CHAPTER 1: INTRODUCTION 6
CHAPTER 2: WHAT IS QUANTUM PHYSICS 7
CHAPTER 3: HEISENBERG UNCERTAINTY PRINCIPLE 8
CHAPTER 4: BLACK BODY 11
CHAPTER 5: PHOTOELECTRIC EFFECT 13
  PHOTOELECTRIC PRINCIPLE 14
  APPLICATIONS 15
CHAPTER 6: THE ATOM 17
  BOHR MODEL 17
  QUANTUM PHYSICS MODEL OF THE ATOM 20
    Wave-particle duality and the de Broglie wavelength 20
    How to calculate the de Broglie wavelength of an electron 21
    The quantum mechanical model of the atom 21
    Schrödinger's equation 21
    Orbitals and probability density 22
    Electron spin: The Stern-Gerlach experiment 24
CHAPTER 7: FRANCK HERTZ EXPERIMENT 25
  EXPERIMENTAL SETUP 25
    Results 26
  WHAT ABOUT THE IMPACT OF ADJUSTING THE TEMPERATURE OF THE TUBE? 28
CHAPTER 8: ENTANGLEMENT 29
CHAPTER 9: PARTICLES CREATION AND ANNIHILATION 37
  ELECTRON–POSITRON ANNIHILATION 37
  PROTON–ANTIPROTON ANNIHILATION 38
CHAPTER 10: QUANTUM TUNNELING 39
  DISCOVERY OF QUANTUM TUNNELING 39
  QUANTUM TUNNELING IN NATURE 39
  TUNNELING IN STELLAR FUSION 39
  TUNNELING IN SMELL RECEPTORS 40
  APPLICATIONS OF QUANTUM TUNNELING 40
    Josephson Junctions 40
    Tunnel Diodes 41
    Scanning Tunneling Microscopes 41
    Flash Drives 41
    Nuclear fusion 42
    Radioactive decay 42
    Astrochemistry in interstellar clouds 42

Quantum biology *43*
Quantum conductivity *43*
Scanning tunneling microscope *43*
Kinetic isotope effect *43*

CHAPTER 11: THE COMPTON EFFECT 45

CHAPTER 12: WAVE-PARTICLE DUALITY 47

CHAPTER 13: SCHRODINGER EQUATION 50

CHAPTER 14: QUANTUM FIELD THEORY 52

CHAPTER 15: PERIODIC TABLE OF ELEMENTS 53

CHAPTER 16: LASERS 57

HISTORY 57
FUNDAMENTAL PRINCIPLES 58
  Energy levels and stimulated emissions *58*
LASER ELEMENTS 59
LASER BEAM CHARACTERISTICS 60
TYPES OF LASERS 61
LASER APPLICATIONS 62
  Transmission and processing of information *62*
  Precise delivery of energy *63*
  Industrial uses *63*
  Medical applications *63*
  High-energy lasers *64*
  Alignment, measurement, and imaging *64*
  Research tool *65*

CHAPTER 17: LED 67

CHAPTER 18: QUANTUM COMPUTING 68

WHAT IS QUANTUM COMPUTING? 68
HOW DO QUANTUM COMPUTERS WORK? 68
WHAT CAN QUANTUM COMPUTERS DO? 69
WHEN WILL I GET A QUANTUM COMPUTER? 69

CHAPTER 19: SUPERCONDUCTIVITY 71

CHAPTER 20: SUPERFLUIDS 74

CHAPTER 21: QUANTUM DOTS 76

THE APPLICATION IN MEDICAL DIAGNOSTICS 76
PROBLEMS WITH QUANTUM DOTS 77

CHAPTER 22: MRI 79

CHAPTER 23: BONUS CHAPTER: RELATIVITY 81

CONCLUSIONS 84

CHAPTER 1: INTRODUCTION

Quantum physics is often thought to be tough because it's related to the study of physics on an unbelievably small scale, though it applies directly to many macroscopic systems. We use the descriptor "quantum" because, as we will see later in the book, when we talk about discrete quantities of energies or packets, in contrast with the classical continuum of Newtonian mechanics.

However, some quantities still attack continuous values, as we'll also see. In quantum physics, particles have wavy properties, and the Schrodinger wave equation, which is a selected differential equation, governs how these waves behave.

In a certain sense, we will discover that quantum physics is another example of a system governed by the wave equation.

It's relatively simple to handle the particular waves. Anyways, we observe several behaviors in quantum physics that are a combination of intricate, enigmatic, and unusual. Some examples are the uncertainty principle that affects measurements, the photoelectric effect, and the wave-particle duality. We will discover all of them in the book.

Even though there are several confusing things about quantum physics, the good news is that it's comparatively simple to apply quantum physics to a physical system to figure out how it behaves. We can use quantum mechanics even if we cannot perceive all of the subtleties of quantum physics.

However, we will excuse ourselves for this lack of understanding of quantum physics because almost nobody understands it (well, maybe a few folks do). If everybody waited until quantum mechanics was fully understood before using it, then we'd be back to the 1920s. The bottom line is that quantum physics works to build predictions that are consistent with experiments.

So far, there are no failures when using it, so it would be foolish not to take advantage of it.

There are so many daily objects we use that are based on technologies derived from quantum mechanics principles: Lasers, Telecommunication (devices and infrastructures), solar energy production, semiconductors, and superconductors, superfluids, medical devices, and much more.

This book will describe quantum mechanics and its experiments and applications. We will try to make it simpler and easy to understand without the complicated math behind it (we will show some formulas and equations).

So, follow me, and let's browse through the **"magic"** world of quantum physics.

CHAPTER 2:

WHAT IS QUANTUM PHYSICS

We define quantum mechanics as that branch of physics that allows the calculation of the probabilities of the "microscopic" world's behavior. This can end up in what might seem to be some strange conclusions concerning the physical world. At the dimensions of atoms and electrons, several classical mechanics equations that describe the motion and speed of the macroscopic world do not work. In classical mechanics, objects exist in an exceedingly specific place at a particular time. However, in quantum physics, particles and sub-particles instead are in a haze of probability; they have an exact possibility of being in a state A and another chance of being in a state B at the same time.

Quantum mechanics developed over several decades, starting as a group of disputable mathematical explanations of experiments that the mathematics of classical mechanics couldn't explain.

It began in the twentieth century when Albert Einstein revealed his theory of relativity, a separate mathematical and physics revolution to describe the motion of objects at speeds close to light speed. Unlike the theory of relativity, however, the origins of QM cannot be attributed to any person. Instead, multiple physicists contributed to a foundation of 3 revolutionary principles that increased acceptance and experimental verification between 1900 and 1930. They are: **Quantized properties:** some properties, like position, speed, and color, will typically solely occur in specific set amounts, very similar to a dial that "clicks" from variety to variety. This observation challenged a fundamental assumption of classical mechanics that states that we can distribute such properties on a continuous spectrum. To explain the behavior where some properties "clicked" like a dial with specific settings, scientists started to use the word "quantized."

Particles of light: we can sometimes describe light behavior as a particle. This light "as a particle" at first raised many criticisms between scientists because it ran contrary to two hundred years of experiments, as it shows that light behaved as a wave, very similar to ripples on the surface of a peaceful lake. Light behaves the same way: it bounces off walls and bends around corners, the crests and troughs of the wave will add up or annul. The sum of more wave crests leads to brighter light, whereas waves that annul themselves turn into darkness. We can imagine a light source like a ball on a stick being rhythmically dipped within the center of a lake. The color emitted corresponds to the gap between the crests; the ball's rhythm's speed decides the color.

Waves of matter: Matter can even behave as a wave. This behavior is counterintuitive as all the known experiments until recent years had shown that matter (such as electrons) exists as particles.

CHAPTER 3:

HEISENBERG UNCERTAINTY PRINCIPLE

In this chapter, we will consider a pillar of quantum physics that places essential constraints on what we can truly measure, a fundamental property first discovered by Werner Heisenberg, referred to as the "Heisenberg Uncertainty Principle."

We are accustomed to thinking that the word "principle" is associated with something that rules the universe. Therefore, when we face the term "Uncertainty Principle," we are disoriented because it looks like juxtaposing opposites. However, in the beginning, the Uncertainty Principle may be a fundamental property of quantum physics, discovered through classical mechanics reasoning (a classically primarily based logic still used by several physics academics to teach the Uncertainty principle these days). This Newtonian approach is that if one examines elementary particles using light, the act of striking the particle with light (even only one photon) will change the particle position and Momentum. It would be impossible to establish where the particle indeed is because now it's not where it was anymore. I know it can look strange. The fact is that at the size level of the elementary particles, the energy (the light) used to make the measurement is of the same order of magnitude as the object you are going to measure. So, it can significantly change the properties of the measured object.

Smaller wavelength lightweight (purple, for instance, that is very energetic) confers a lot of energy to the particle than longer-wavelength light (red is less energetic). Therefore, employing a smaller (more precise) "yardstick" of light to measure position means one interferes a lot with the potential position of the particle by "hitting" it with more energy. Whereas his sponsor, Niels Bohr (who with success argued with Einstein on several of those matters), was on travel, Werner Heisenberg, first printed his uncertainty principle paper using this more or less classical reasoning. (The deviation from classical notion was that light comes in packets or quantities, referred to as "quanta"). Heisenberg, after his first paper, never imagined that the uncertainty principle would become even more fundamental.

The concept of Momentum is fundamental in physics. In classical mechanics, we define it as the mass of a particle multiplied by its velocity. We can imagine a baseball thrown at us at a hundred miles per hour having an identical result as a bat thrown at us at 10 miles per hour; they'd each have similar Momentum though they have a completely different mass.

The Heisenberg Uncertainty Principle states that the moment we start to know the variation in the Momentum of an elementary particle accurately (that means we measure the particle's velocity), we begin to lose knowledge about the modification of the position of the particle (that is, where the particle stays). Another way of phrasing this principle is using relativity in the formulation. This relativistic version states that as we get to know the energy of an element accurately, we cannot at the same time know (i.e., measure) some right information such as what time it had that energy. So, we have, in quantum physics, the "complementary pairs." (Or, if you'd like to stand out with your companions, "non-commuting observables.")

We can show the results of the uncertainty principle with a not-filled balloon. We could write "delta-E" on the right side to represent our particle energy uncertainty value. On the left side of the balloon, write "delta-t," which would stand for when the particle had that energy uncertainty. When we compress the delta-E side (so that it fits into our hand), we can observe that the delta-t side would increase its size. Similarly, if we want the delta-t side to stay within our hand, the delta-E side would increase its volume. In this example, the cumulative value of air in the balloon would remain constant; it would just shift its position. The quantity of air in the balloon is one quantity, or one "quanta," the tiniest unit of energy viable in quantum physics. We can append more quanta-air to the balloon (making all the values higher, both in delta-E and delta-t), but we can never take more than one quanta-air out of the balloon in our case. Thus "quantum balloons" do not arrive in packets any less than one quanta or photon.

When quantum mechanics was still in the early stages, Albert Einstein (and colleagues) would debate with Niels Bohr team about many unusual quantum puzzles. Some of these showed the possibility that the results would suggest that elementary particles, by quantum effects, could interact faster than light. Light speed infraction is the main reason Einstein was known to imply that since we have outcomes that deny the speed-of-light limit set by relativity, we could not understand how these physics effects could occur. Einstein suggested various experiments one could perform, the most notable being the EPR (Einstein, Podolski, Rosen) paradox, which showed that a faster than light connection would seem to be the outcome. Consequently, he argued that quantum mechanics was unfinished and that some factors had to be undiscovered. Einstein considerations motivated Niels Bohr and his associates to express the "Copenhagen Interpretation" of quantum physics reality. This interpretation is incorrect to consider as an elementary particle doesn't exist unless observed. Rephrasing the concept: elementary particles should be regarded as being made up of forces, but the observer is also a component to consider. The observer can never really be segregated from the observation.

Using Erwin Schrödinger wave equations for quantum particles, Max Born was the first to suggest that these elementary particle waves were made up of probabilities. So, the ingredients of everything we see are composed of what we might call "tendencies to exist," which are built into particles by adding the necessary component of "looking." There are other possible explanations we could follow. We can say that none of them was consistent with any sort of objective reality as Victorian physics had comprehended it before. Other theories could fit the data well.

All of these theories have one of these problems:

1) They are underlying faster than light transmission (theory of David Bohm).

2) There is another parallel universe branching off ours every time there is a small decision to be made.

3) The observer produces the reality when he observes (the Copenhagen Interpretation).

Excited by all these theories, a physicist at CERN in Switzerland named John Bell set up an experiment that would test some of the theories and examine how far quantum physics was from classical physics. By then (1964), quantum mechanics was old enough to have defined itself from all previous physics. Then physics before 1900 was called "classical physics," and physics discovered after 1900 was tagged "modern physics." So, science was broken up in the first 46 centuries (if we consider Imhotep, who developed the first pyramid, as the first historical scientist) and the last century, with quantum physics. So, we can say that we are relatively young when it comes to modern physics, the new fundamental view of science. We may say that most people are not even conscious of the development that occurred in the fundamental basis of the scientific endeavor and interpretations of reality, even after a century.

John Bell planned an experiment that could evaluate two given elementary particles. He proposed that if they were farther away from each other, they could "communicate" between themselves faster than when any light traveled between them. In 1984, in Paris, Alain Aspect led a team to perform this experiment. He executed the experiment using polarized light. Let's consider a light container, where the light is waving all over the place. If the container is covered with a reflective material (except for the ends), the light bounces off the walls. At the termination, we place polarizing filters, which means that only light with a given orientation can get out, while back-and-forth light waves cannot get out. If we twist the polarizers at both edges by 90 degrees, we would then let out back-and-forth light waves, but now not up-and-down light.

If we were to twist the terminations at an angle of 30 degrees to each other, it happens that about half of the total light could escape the container -- one-fourth from one side of it and one-fourth through the other side. This experiment is what John Bell proposed, and Alain Aspect confirmed. When the container was turned at one edge, making a 30-degree angle with the other end so that only half the light could appear, a shocking thing occurred. After turning one side of the container, the light was coming out of the opposite surface immediately (or as close to immediate as anyone could measure) before any light could reach the other side of the container. Somehow the message, that one end had been turned traveled faster than the speed of light. Since then, this experiment has been replicated many times with the same result.

John Bell's formulation of the basic ideas in this research has been called "Bell's Theorem" and can be declared most succinctly in his terms: "Reality is non-local." That is to say, not only do the elementary particles not exist until they are observed (Copenhagen Interpretation), but they are not, at the primary level, even identifiably detachable from other particles arbitrarily.

CHAPTER 4:

BLACK BODY

The idea of quantized energies emerged with one of the most known challenges to classical physics: the problem of black-body radiation. While classical physics Wien's formula and the Rayleigh-Jeans Law could not describe the spectrum of a black body, Max Planck's equation explained the enigma by understanding that light was discrete.

When you warm an object, it begins to shine. Even before the brightness is visible, it's radiating in the infrared spectrum. It emits because as we heat it, the electrons on the facade of the material are excited thermally, and accelerated and decelerated electrons diffuse light. Between the end of the 19th and the start of the 20th centuries, physics was concerned with the spectrum of light emitted by black bodies. A black body is a piece of material that radiates in correspondence to its temperature — but most of the everyday objects you think of as black, such as charcoal, also embody and reflect light from their surroundings. Physics assumed that a black body reflected nothing. It absorbed all the light falling on it (hence the term *black body*, because the object would resemble wholly black as it absorbed all light falling on it). When you heat a black body, it will radiate, emitting light.

It was difficult to develop a physical black body because no material absorbs light a hundred percent and reflects anything. But the physicists were smart about this, and they came up with the hollow cavity with a hole in it. When we irradiate light in the hole, all that light will go inside, where it would be reflected again and again — until it got absorbed (only a negligible quantity of light would leave the cavity through the hole). And when we heated the hollow cavity, the hole would begin to radiate. So, we have a good approximation of a black body.

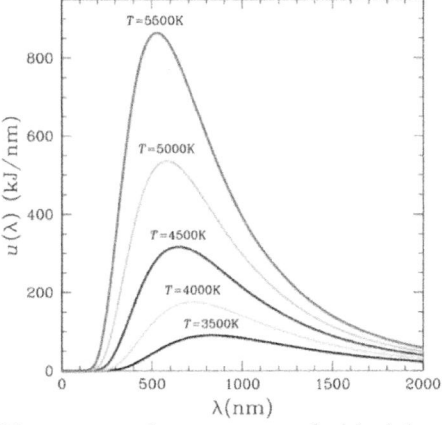

You can see the spectrum of a black body (and tries to model that spectrum) in the above figure for two distinct temperatures, T1 and T2. The difficulty was that nobody was able to come up with a theoretical solution for the spectrum of light formed by the black body. Everything classical physics could try went wrong.

The black-body enigma was a difficult one to solve, and with it came the origins of quantum physics. Max Planck proposed that the quantity of energy that a light wave can exchange with the matter wasn't continuous but discrete, as hypothesized by classical physics. Max Planck presumed that the energy of the light released from the walls of the black-body cavity came exclusively in integer multiples like this, where h is a universal constant, E is the energy, n is an integer number, and ν is the frequency:

E=nhν

With this theory, absurd as it appeared in the early 1900s, Planck converted the continuous integrals used by Raleigh-Jeans in the classical mechanics to discrete sums across an infinite number of terms. Making that simple change gave Planck the following equation for the spectrum of black-body radiation:

$$B_\nu(\nu, T) = \frac{2h\nu^3}{c^2} \frac{1}{e^{h\nu/kT} - 1}$$

Where:
- $B\nu(T)$: spectral radiance density of frequency ν radiation at thermal equilibrium at temperature T.
- h: Planck constant.
- c: Speed of light in a vacuum.
- k: Boltzmann constant.
- ν: Frequency of the electromagnetic radiation.
- T: Absolute temperature of the body.

This equation was right — it accurately defines the black-body spectrum, both at low and high (and medium, for that matter) frequencies.

This idea was new. Planck was saying that the energy of the radiating oscillators in the black body couldn't work on any level of energy; it could consider only precise, quantized energies.

Planck's hypothesis was true for any oscillator — that its energy was an integer multiple of hν.

This is the *Planck's quantization rule*, and h is Planck's constant:
- h = 6.626*10-34 Joule-seconds

Discovering that the energy of all oscillators was quantized gave birth to quantum physics.

CHAPTER 5:

PHOTOELECTRIC EFFECT

When a material absorbs electromagnetic radiation, and as a consequence, releases electrically charged particles, we observe the photoelectric effect. We usually think of the photoelectric effect as the ejection of electrons from a metal plate when light falls on it. But the concept can be extended: the radiant energy may be infrared, visible, ultraviolet light, X-rays, gamma rays; the material may be a gas, liquid, solid; the effect can release ions (electrically charged atoms or molecules) other than electrons. The phenomenon was practically meaningful in expanding modern physics because of the difficulties related to questions about the nature of light (particle versus wavelike behavior). Albert Einstein finally resolved the problem in 1905 (gaining the Noble prize). The effect remains essential for research in areas ranging from materials science to astrophysics and forms the basis for various useful devices. The first to discover the photoelectric effect in 1887 was the German physicist Heinrich Rudolf Hertz. Hertz was working on radio waves experiments. He recognized that when ultraviolet light irradiates on two metal electrodes with a voltage applied across them, the light alters the sparking voltage. In 1902, Philipp Lenard, a German physicist, clarifies the relation between light and electricity (hence photoelectric). He showed that electrically charged particles are released from a metal surface when it is lighted. These particles are equal to electrons, which had been found by the British physicist Joseph John Thomson in 1897. Additional research revealed that the photoelectric effect denotes an interaction between light and matter that classical physics cannot explain; it describes light as an electromagnetic wave. One puzzling observation was that the maximum kinetic energy of the freed electrons did not vary with the intensity of the light, as presumed by the wave theory, but was comparable instead to the frequency of the light. The light intensity determined the number of electrons released from the metal (measured as an electric current). Another difficult observation was that there was virtually no time delay between the arrival of radiation and the emission of electrons. Einstein's model explained the emission of electrons from a lighted plate; anyway, his photon theory was so radical that the scientific community did not wholly trust until it got further experimental confirmation. In 1916, Robert Millikan (an American physicist) gave additional corroboration whit as highly accurate measurements that confirmed Einstein's equation and exposed with high precision that Einstein's constant h was the same as Planck's constant. Einstein finally received the Nobel Prize for Physics in 1921 for describing the photoelectric effect. In 1922 Arthur Compton (an American physicist) measured the shift in wavelength of X-rays in case of interaction with free electrons, and he showed that we could calculate the change by handling X-rays as if they're made of photons. Compton won the 1927 Nobel Prize for Physics for this work. Ralph Howard Fowler, a British mathematician, extended our understanding of photoelectric emission in 1931 by proving the relationship between temperature in metals and photoelectric current. Further studies revealed that electrons could be released by electromagnetic radiation in insulators, which do not conduct electricity. In semiconductors, a variety of insulators conduct electricity only under certain circumstances.

Photoelectric Principle

According to quantum physics, electrons bound to atoms occur in precise electronic configurations. The valence band is the highest energy configuration (also called energy band) that is usually occupied by electrons of a given element. The degree to which it is loaded determines the material's electrical conductivity. In a regular conductor (metal), the valence band is about half-filled with electrons, which promptly move from atom to atom, conducting a current. In a good insulator, such as plastic, the valence band is filled, and these valence electrons have very little fluidity. Semiconductors, generally, have their valence bands congested. Still, unlike insulators, very little energy can excite an electron to the next allowed energy band (known as the conduction band) because any electron excited to this higher energy level is nearly free. Let's consider a semiconductor like silicon: its "bandgap" is 1.12 eV (electron volts). This value is compatible with the energy carried by photons of infrared and visible light. This small amount of energy can lift electrons in semiconductors to the conduction band. Depending on the semiconducting material configuration, this radiation may enhance its electrical conductivity by adding to an electric drift. Applied voltage (photoconductivity) or any external voltage sources (photovoltaic effect) can induce electric drift. When light frees the electrons or a flow of positive charge, we can see photoconductivity. When excited, electrons surge to the conduction band in the valence band, and we can observe the formation of negative charges, called "holes." If we illuminate the semiconductor material, both electrons and holes increase current flow.

On the other side, when we are in the presence of the photovoltaic effect, we can observe that a voltage is generated when the electrons liberated by the incident light are split from the holes that are generated, causing a difference in electrical potential. We can see this happening by using a p-n junction or a semiconductor. At the juncture between n-type (negative) and p-type (positive) semiconductors, we have a p-n junction. Generation of different impurities creates these opposite regions excess electrons (n-type) or excess holes (p-type). Light releases electrons through holes on opposite sides of the junction to provide a voltage across the junction that can drive current, thereby turning light into electrical power.

Radiation at higher frequencies, like X-rays and gamma rays, causes other photoelectric effects. These higher-energy photons can provide enough energy to release electrons close to the atomic nucleus, where they are tightly bound. In this case, when the photon collision ejects an inner electron, a higher-energy outer electron quickly drops down to fill the vacancy. The difference in energy results in the ejection of one or more further electrons from the atom: we have described the Auger effect. We can also phase the Compton Effect when we use high photon energies, which arises when an X-ray or gamma-ray photon strikes with an electron. We can analyze the effect with the same principles that govern the collision between two bodies, including preserving momentum. The photon expends energy (gained by the electron), a reduction that corresponds to an enlarged photon wavelength according to Einstein's relation $E = hc/\lambda$.

When the electron and the photon are at right angles to each other after the collision, the photon's wavelength increases by a characteristic amount called the Compton wavelength, 2.43×10^{-12} meter.

Applications

Devices based on the photoelectric effect have various beneficial characteristics, including the ability to produce a current proportional to light intensity and a quick response time. One of such devices is the photoelectric cell or photodiode. This was formerly a phototube, a vacuum tube including a cathode made of metal, and this has a small work function; thus, electrons would be easily transmitted. The current delivered by the plate would be collected by an anode and kept at a large positive voltage compared to the cathode. Phototubes have been supplanted by semiconductor-based photodiodes that can detect light, estimate its intensity, and manage other devices by serving as a function of illumination and converting light into electrical energy. These machines work at low voltages, analogous to their bandgaps. They are utilized in pollution monitoring, industrial process control, solar cells, light detection within fiber optics telecommunications networks, imaging, and many more.

Photoconductive cells are built using semiconductors that possess bandgaps relevant to the photon energies that need to be sensed. For example, automatic switches for street lighting and photographic exposure meters work in the visible spectrum, so they are usually produced with cadmium sulfide. Infrared detectors (like sensors for night-vision purposes) are made of lead sulfide or mercury cadmium telluride. Photovoltaic devices typically include a semiconductor p-n junction. For solar cell use, they usually are made of crystalline silicon and transform a little more than 15 percent of the hitting light energy into electricity. Solar cells are often used to provide relatively small amounts of power in special conditions such as remote telephone installations and space satellites. The development of more affordable and higher efficiency supplies may make solar power economically worthwhile for large-scale applications.

First developed in the 1930s, the photomultiplier tube is an extremely sensitive extension of the phototube. It includes a series of metal plates named dynodes. Light hitting the cathode releases electrons. The electrons are attracted to the first dynode, where they free additional electrons that hit the second dynode, and so on. After up to 10 dynode steps, the photocurrent is enormously increased. Some photomultipliers can detect a single photon. These tools, or solid-state versions of analogous sensitivity, are priceless in spectroscopy research, where it is frequently needed to measure bare light sources. Also, scintillation counters utilize photomultiplier. These scintillation counters contain a substance that creates flashes of light when struck by gamma rays or X-rays. It is joined to a photomultiplier that calculates the flashes and measures their intensity. These counters help applications identify particular isotopes for nuclear tracer analysis and detect X-rays used in computerized axial tomography (CAT) scans to represent a cross-section through the body.

Imaging technology also uses photodiodes and photomultipliers. Television camera tubes, light amplifiers or image intensifiers, and image-storage tubes use the fact that the electron emanation from each point on a cathode is defined by the number of photons reaching that point. An optical image coming on one side of a semitransparent cathode is turned into an equivalent "electron current" image on the other side. Then electric and magnetic fields are applied to sharpen the electrons onto a phosphor screen. Each electron hitting the phosphor generates a flash of light, provoking the freedom of more and more electrons from the relevant point on a cathode opposite the phosphor. It is possible to considerably enhance the resulting amplified image by offering even greater amplification and being represented or saved.

At more important photon energies, the analysis of electrons transmitted by X-rays provides knowledge about electronic transitions among energy states in molecules and atoms. It also adds to the study of specific nuclear processes. It plays a part in the chemical analysis of elements since emitted electrons give specific energy exclusive to the atomic source. The Compton Effect is also applied to analyze the properties of materials, and in astronomy, it is also used to investigate gamma rays that come from cosmic sources.

CHAPTER 6:

THE ATOM

Bohr Model

Danish physicist Niels Bohr was engaged in understanding why the observed spectrum was composed of discrete lines when the different elements emitted light. Bohr was involved in defining the structure of the atom. This topic was of much interest to the scientific community. Various models of the atom had been proposed based on empirical results. Some of them included the electron detection by J. J. Thomson and the nucleus's acknowledgment by Ernest Rutherford. Bohr recommended the planetary model, in which electrons rotated around a positively charged nucleus like planets around the star.

However, scientists still had multiple pending problems:
- What is the position of the electrons, and what do they do?
- Why don't electrons collapse into the nucleus as prognosticated by classical physics while rotating around the nucleus?
- How is the inner structure of the atom linked to the discrete radiation lines produced by excited elements?

Bohr approached these topics using a simple assumption: what if some properties of atomic structure, such as electron orbits and energies, could only consider specific values? By the early 1900s, scientists were conscious that some events occurred in a discrete, as opposed to a continuous manner. Physicists Max Planck and Albert Einstein had recently hypothesized that electromagnetic radiation not only acts like a wave but also sometimes like particles named *photons*. Planck studied the electromagnetic radiation emitted by heated objects, and he proposed that the emitted electromagnetic radiation was "quantized" since the energy of light could only have values given by the following equation:

$E = nh\nu$

Where n is an integer (positive), h is Planck's constant 6.626×10^{-34} Joule seconds, ν is the frequency of the light.

As a consequence, the emitted electromagnetic radiation must have energies that are multiples of $h\nu$. Einstein used Planck's results to justify why we need a minimum frequency of light to expel electrons from a metal exterior in the photoelectric effect.

When something is **quantized**, it implies that we can consider only distinct values, such as when playing the piano. Since each key of a piano is tuned to a definite note, just a specific set of notes (relevant to frequencies of sound waves) can be emitted. As long as your piano is accurately tuned, you can play an F or F sharp, but you can't play the note between an F and F sharp.

Atomic line spectra are another example of quantization. When we heat an element or ion by a flame or excite it by an electric current, the excited atoms emit light of a specific color. A prism can refract the emitted light, producing spectra with a distinctive striped appearance due to the radiation of particular wavelengths of light. When we consider the hydrogen atom, we can use mathematical equations to fit the wavelengths of some emission lines, but the equations do not explain why the hydrogen atom emitted those particular wavelengths of light. Before Bohr's model of the hydrogen atom, physicists were unaware of the motivation behind the quantization of atomic emission spectra.

Bohr's model of the hydrogen atom originated from the planetary model, but he combined one assumption concerning the electrons. He wondered if he could consider the electronic structure of the atom as quantized. Bohr proposed that the electrons could only orbit the nucleus in distinct orbits or *shells* with a fixed radius. He also wrote an equation where just shells with a given radius would be allowed, and the electron cannot stay in between these shells. Mathematically, we could write the permitted values of the atomic radius as $r(n)=n^2*r(1)$ where n is a positive integer, and $r(1)$ is the **Bohr radius**, the smallest allowed radius for hydrogen:

Bohr radius = $r(1) = 0.529 \times 10^{-10}$m

By keeping the electrons in circular, and quantized orbits around the positively-charged nucleus, Bohr could calculate the energy of any level of an electron in the hydrogen atom: $E(n)=-1/n^2 * 13.6eV$ where the lowest energy or ground state energy of a hydrogen electron $E(1)$ is -13.6eV.

Note that the energy will always be a negative number, and the ground state, n=1, equals 1, has the most negative value. When we separate an electron from its nucleus (n=∞), the defined energy is 0 eV: for this reason, the energy of an electron in orbit is negative. An electron in orbit always has negative energy because an electron in orbit around the nucleus is more stable than an infinitely distant electron from its nucleus.

Quantum Physics for Beginners

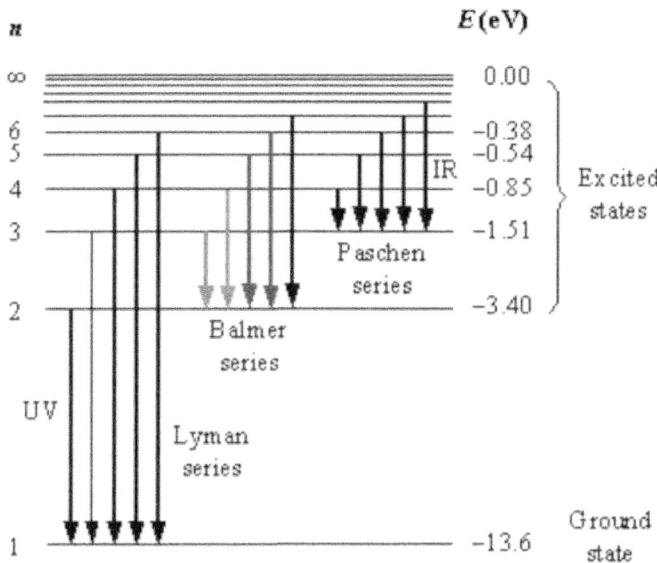

Bohr had explained the processes of absorption and emission in terms of electronic structure. According to Bohr's model, an electron gets excited to a higher energy level when absorbing energy in the form of photons *as long as the photon's energy was equal to the energy gap between the original and final energy levels.* The electron would be in a less stable position after jumping to the higher energy level (or **excited state),** so it would quickly emit a photon in order to be back to a lower, more stable energy level. We can illustrate the energy levels and transitions between them using an **energy level diagram.** The difference in energy between the two energy levels for a particular transition may be used to calculate the photon emitted energy.

The energy difference between energy levels:

$$\Delta E = \left(\frac{1}{n_{low}^2} - \frac{1}{n_{high}^2} \right) \cdot 13.6\,\text{eV}$$

We can also identify the equation for the wavelength and the frequency of the emitted electromagnetic radiation. We use the relationship between the speed of light frequency and wavelength and the relation between energy and frequency.

$$\nu = \left(\frac{1}{n_{low}^2} - \frac{1}{n_{high}^2} \right) \cdot \frac{13.6\,\text{eV}}{h}$$

$$\frac{1}{\lambda} = \left(\frac{1}{n_{low}^2} - \frac{1}{n_{high}^2} \right) \cdot \frac{13.6\,\text{eV}}{hc}$$

Therefore, the emitted photon's frequency (and wavelength) depends on the difference of the initial and final energies of the shells of an electron in hydrogen.

The Bohr model worked to explain the hydrogen atom and other single-electron systems such as He^+. Unfortunately, the model could not be applied to more complex atom structures. Furthermore, the Bohr model could not explain why some spectral lines split into multiple lines when experiencing a magnetic field (the Zeeman Effect) or why some are more intense than others.

In the following years, other physicists such as Erwin Schrödinger determined that electrons behave like waves and behave as particles. The result was that we could not know an electron's position in space and its velocity at the same time (a concept more accurately stated in *Heisenberg's uncertainty principle*). The uncertainty principle denies Bohr's conception of electrons existing in particular orbits with an identified velocity and radius. We can calculate *the probabilities* of locating electrons in a specific region of space around the nucleus.

The modern quantum physics model may appear like a massive leap from the Bohr model. The fundamental concept is the same: classical physics doesn't describe all phenomena on an atomic level. Bohr was the first to acknowledge this by including the idea of quantization into the hydrogen atom's electronic structure. He was able to explain the radiation spectra of hydrogen and other one-electron systems.

Quantum Physics Model of the Atom

At the subatomic level, matter begins to behave very strangely. We can only talk about this counterintuitive behavior with symbols and metaphors. For example, what does it mean to state that an electron behaves like a wave and a particle? Or that we need to think of an electron as not existing in any location. But we have to imagine it as being *spread out* entirely in the entire atom.

If these inquiries make no sense to you, it is normal, and you are in good company. The physicist Niels Bohr said, "Anyone who is not shocked by quantum theory has not understood it." So, if you feel disoriented when learning about quantum physics, know that the physicists who initially worked on it were just as muddled.

While some physicists attempted to adapt Bohr's model to explain more complicated systems, others decided to study a radically different model.

Wave-particle duality and the de Broglie wavelength

French physicist Louis de Broglie pioneered another significant development in quantum physics. Based on Einstein and Planck's work that explained how light waves could display particle-like properties, de Broglie thought that particles could also have wavelike properties.

De Broglie obtained the equation for the wavelength of a particle of mass m traveling at velocity v (in m/s), where λ is the de Broglie wavelength of the particle in meters and h is Planck's constant:

$\lambda = h/mv$

The particle mass and the de Broglie wavelength are inversely proportional. We don't observe any wavelike behavior for the macroscopic objects we encounter every day because of the inversely proportional relationship between mass and wavelength. Consequently, the wavelike behavior of matter is most notable when a wave meets an obstacle or slit similar to its de Broglie wavelength. When the mass of a particle is on the order of 10^{-31} kg, as an electron does, it starts to show the wavelike behavior, leading to some fascinating phenomena.

How to calculate the de Broglie wavelength of an electron

The velocity of an electron in the ground-state energy level of hydrogen is $2.2*10^6$ m/s. If the electron's mass is $9.1*10^{-31}$ kg, its de Broglie wavelength is $\lambda=3.3*10^{-10}$ meters, is on the same order of magnitude as the diameter of a hydrogen atom, $\sim 1*10^{-10}$ meters. That means an electron will often encounter things with a similar size as its de Broglie wavelength (like a neutron or atom). When that happens, the electron will show wavelike behavior!

The quantum mechanical model of the atom

Bohr's model's major problem was that it managed electrons as particles that existed in precisely defined orbits. Based on de Broglie's idea that particles could display wavelike behavior, Austrian physicist Erwin Schrödinger hypothesized that we could explain the behavior of electrons within atoms, considering them mathematically as matter waves. Based on this modern understanding of the atom, we call this model the *quantum mechanical* or *wave mechanical* model.

An electron in an atom can have specific admissible states or energies like a *standing wave*. Let's briefly discuss some properties of standing waves to get a more solid intuition for electron matter waves.

You are probably already familiar with standing waves from stringed musical instruments. For example, when we pluck a string on a guitar, the string vibrates in the shape of a standing wave.

Notice that there are points of zero displacement, or *nodes* that occur along with the standing wave. The string (that is attached at both ends) allows only specific wavelengths for any standing wave. As such, the vibrations are quantized.

Schrödinger's equation

Let's see how to relate standing waves to electrons in an atom.

We can think of electrons as standing matter waves that have specifically allowed energies. Schrödinger formulated a model of the atom where he could treat electrons as matter waves. The primary form of Schrödinger's wave equation is: $\hat{H}\psi=E\psi$, ψ(psi) is a *wave function*; $\hat{H}$ is the Hamiltonian operator, and E is the binding energy of the electron. The result of Schrödinger's equation is multiple wave functions, each with an allowed value for E.

Interpreting what the wave functions tell us is a bit tricky. According to the *Heisenberg uncertainty principle*, we cannot identify both the energy and position of a given electron.

Since identifying the energy of an electron is essential for predicting the chemical reactivity of an atom, chemists accept that we can only approximate the position of the electron. How can we approximate the position of the electron? We call atomic orbitals the Schrödinger's equation wave functions for a specific atom. *There is a region within an atom where the probability of finding an electron is 90% of the time: chemists call this region an atomic orbital.*

Orbitals and probability density

The value of the wave function ψ at a given point in space (x, y, z) is proportional to the electron matter wave's amplitude at that point. However, many wave functions are complex functions containing $i=\sqrt{-1}$, and the matter wave's amplitude has no real physical significance.

Luckily, the *square* of the wave function, ψ^2, is a little more useful. The *probability* of finding an electron in a particular volume of space within an atom is proportional to the square of a wave function. The function ψ^2 is often called the *probability density*. We can visualize the probability density for an electron in several different ways. For example, we can represent ψ^2 by a graph in which we can change the intensity of the color to show the relative probabilities of finding an electron in an assigned region in space. We give a higher density of the color to the region where we have a greater probability of finding an electron in a particular volume. As in the picture, we can represent the probability distributions for the spherical 1s, 2s, and 3s orbitals.

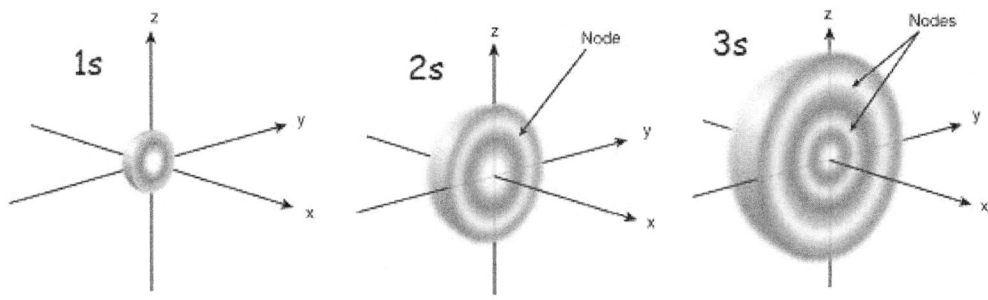

The 2s and 3s orbitals contain nodes—regions where we have a 0% probability of finding an electron. The presence of nodes is similar to the standing waves previously discussed. The alternating colors in the 2s and 3s orbitals delineate the orbital regions with different phases, an essential consideration in chemical bonding. Another way of drawing probability for electrons in orbitals is by plotting the surface density as a function of the gap from the nucleus, r.

Quantum Physics for Beginners

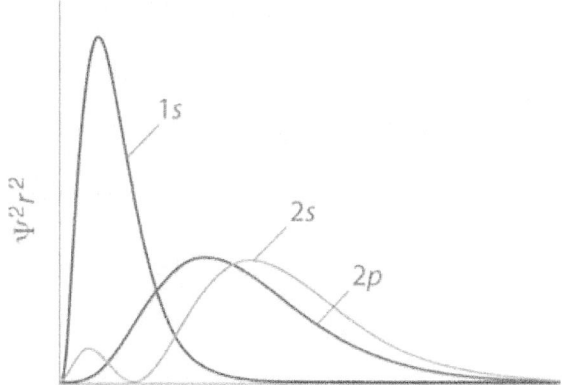

Distance from nucleus (r)

The surface density is the probability of finding the electron in a tiny shell with radius r. We call it a *radial probability* graph. On the left is a radial probability graph for the 1s, 2s, and 3s orbitals. If we increase the energy level of the orbital from 1s to 2s to 3s, the probability of finding an electron farther from the nucleus increases.

So far, we have been examining spherical s orbitals. The gap from the nucleus r is the main factor affecting an electron's probability distribution. Nonetheless, for other types of orbitals such as p, d, and f orbitals, the electron's angular position corresponding to the nucleus becomes a factor in the probability density: it leads to more interesting orbital shapes, like the ones in the following image.

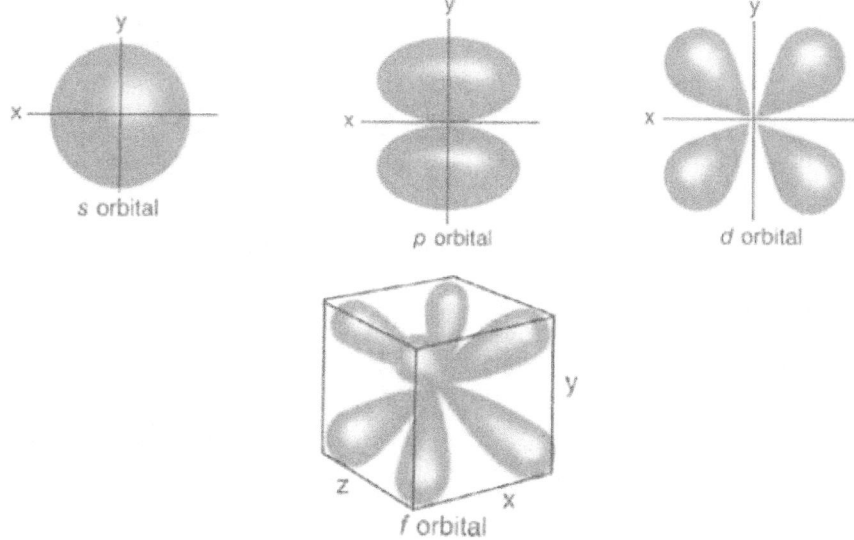

The shape of the p orbitals is like dumbbells oriented along one of the axes (x, y, z). We can describe the d orbitals as having a clover shape with four possible orientations (except for the d orbital that almost seems like a p orbital with a donut going around the middle). The f orbitals are too complicated to describe!

Electron spin: The Stern-Gerlach experiment

In 1922, German physicists Walther Gerlach and Otto Stern hypothesized that electrons acted as little bar magnets, each with a north and south pole. To examine this theory, they shot a beam of silver atoms between the poles of a permanent magnet having a stronger north pole than the south pole.

According to classical physics, the orientation of a dipole in an external magnetic field should define the direction in which the beam gets deflected. Since a bar magnet can have a range of orientations corresponding to the external magnetic field, they expected to see atoms deflect by different amounts to give a spread-out distribution. Instead, Stern and Gerlach observed a clean split of the atoms between the north and south poles.

These experimental results revealed that electrons could exhibit two possible orientations, unlike regular bar magnets: either with the magnetic field or against it. This phenomenon, where electrons can be in one of two possible magnetic states, could not be explained using classical physics! Physicists refer to this property of electrons as *electron spin*: any given electron is either spin-up or spin-down. We sometimes describe electron spin by drawing electrons as arrows pointing down↓ or up ↑.

One outcome of electron spin is that a maximum of two electrons can utilize any given orbital. The two electrons filling the same orbital must have opposite spin: **Pauli Exclusion Principle**.

CHAPTER 7:

FRANCK HERTZ EXPERIMENT

We are at the beginning of the 20th century, and quantum theory was in its infancy. As we saw in previous chapters, the fundamental principle of this quantum world was quantized energy. In other words, we can think of light as being made up of photons, each carrying a unit (or quanta) of energy, and that electrons can only stay in discrete energy levels within an atom.

The Franck-Hertz experiment, conducted by James Franck and Gustav Hertz, was performed in 1914, and it confirmed these discretized energy levels for the first time. It was a historical experiment, rewarded with the 1925 Nobel Prize in Physics. Albert Einstein had this to say about the experiment: *"It's so lovely, it makes you cry!"*

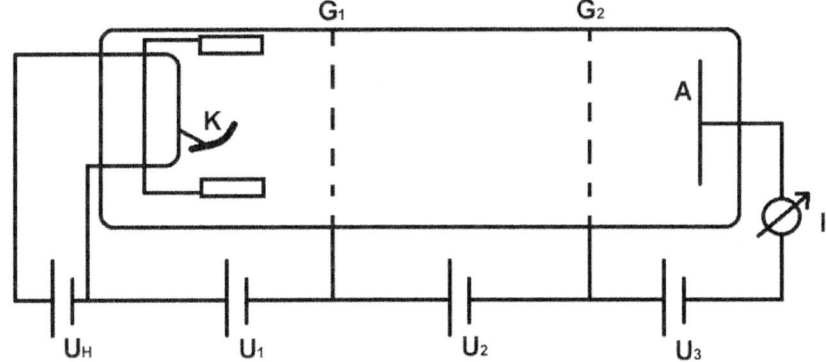

Experimental Setup

To execute the experiment, Franck-Hertz used a tube (shown in the picture). They applied pumps to the tube to evacuate it and form the vacuum. Then they filled the tube with an inert gas (like mercury or neon).

They maintained a low pressure in the tube low and kept a constant temperature. They used a control system for the temperature so that they could adjust it when needed.

The experiment execution measured the current **I** by collecting the output via an oscilloscope or a graph plotting machine.

They applied four different voltages across different segments of the tube. Let's describe the tube sections (starting from left to right) to understand it and how it produced current. The first voltage, **UH**, is responsible for heating a metal filament, **K**. This generates free electrons via thermionic emission (in this way, the heat energy uses the electrons' work function to separate the electron from its atom).

There is a metal grid, G1, near the filament. The grid voltage fixed value is **V1**. This voltage is applied to bring the newly free electrons, which then cross the grid.

Quantum Physics for Beginners

An accelerating voltage, **U2**, is then utilized.

This voltage accelerates the electrons towards the following grid, **G2**. This second grid stopping voltage value is **U3**, which acts to resist the electrons arriving at the collecting anode, **A**.

The electrons gathered at this anode present the measured current. When **UH**, **U1** and **U3** are set, the experiment boils down to changing the accelerating voltage and witnessing the effect on the current.

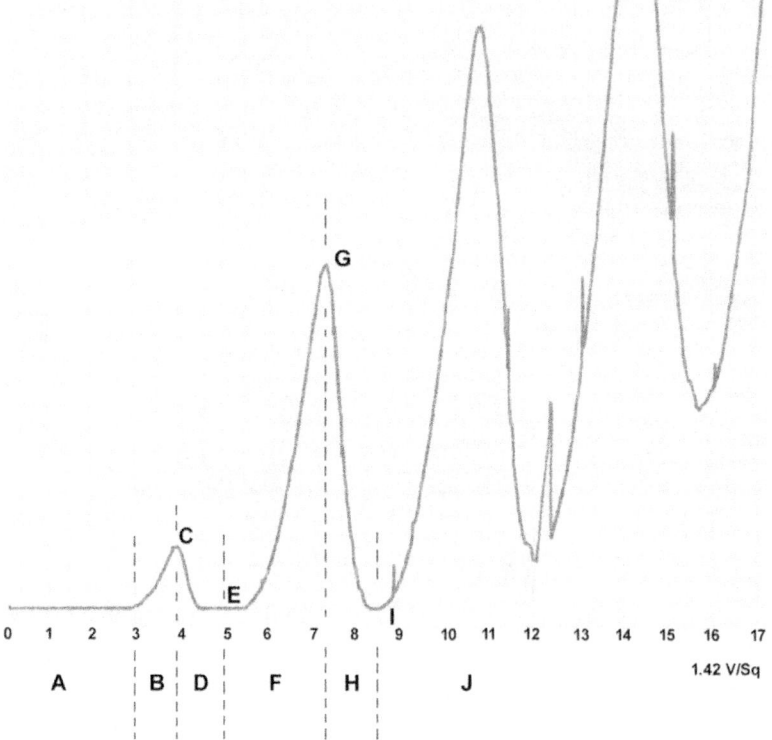

Results

The diagram shows an example of the shape of a typical Franck-Hertz curve. The essential parts got marked with a label on the chart. If we assume that the atom has discretized energy levels, we can find two types of collisions that the electrons can have with the gas atoms in the tube:

- Elastic collisions. It occurs when the electron "bounces" off the gas atom, maintaining its energy/speed. In this case, only the direction of travel changes.
- Inelastic collisions. It occurs when the electron excites the gas atom and dissipates energy. Due to the discrete energy levels, these collisions can only occur for a specific energy value (which is the excitation energy and is relevant to the energy gap between the atomic ground state and a higher energy level).

A - No observed current.

When the accelerating voltage isn't strong enough to triumph over the stopping voltage, then no electrons reach the anode, and there is no current.

B - The current increases to a 1st maximum.

The accelerating voltage gives the electrons enough energy to overcome the stopping voltage but not sufficient to excite the gas atoms. As the acceleration voltage rises, the electrons gain more kinetic energy: the time to cross the tube is less, and consequently, the current increases ($I = Q/t$).

C - The current reaches the 1st maximum.

The accelerating voltage is enough to give electrons energy to excite the inert gas atoms. At this stage, inelastic collisions can begin. After such a collision, the electron may not have sufficient energy to defeat the stopping potential, so the current will decrease.

D - The current set down from the 1st maximum.

Electrons are moving at different speeds and directions because of elastic collisions with the gas atoms (having their random thermal movement). Therefore, some electrons will need more acceleration than others to gain the excitation energy. As a consequence, the current drops instead of declining sharply.

E - The current reaches the 1st minimum.

The collisions exciting the gas reach their maximum number. Hence, almost all electrons are not getting the anode, and there is a minimum current.

F - The current increases again, up to a 2nd maximum.

The accelerating voltage rises enough to accelerate electrons to overcome the stopping potential after they have suffered an energy drop due to an inelastic collision. The average location of inelastic collisions shifts leftwards down the tube, closer to the filament. The current arises due to the kinetic energy discussion outlined previously in **B**.

G - The current reaches the 2nd maximum.

There is enough accelerating voltage to give electrons enough energy to excite two gas atoms while it progresses through the length of the tube. The electron is quickened, has an inelastic collision, accelerated again, has an extra inelastic collision, and then doesn't have adequate energy to overcome the stopping potential, so the current starts to drop.

H - The current drops from the 2nd maximum.

The current constantly drops, as already described previously in **D**.

I - The current reaches the 2nd minimum.

We reach a maximum number of inelastic collisions between electrons and gas atoms. Therefore, many electrons are not hitting the anode, and we observe a second minimum current.

J - The pattern of maxima and minima then replicates for higher and higher accelerating voltages.

The pattern then reappears as more and more inelastic collisions are fitted into the length of the tube.

We can see that the minima of the Franck-Hertz curves are uniformly separated (excluding experimental contingencies). This separation of the minima is equal to the gas atoms' excitation energy (in the case of mercury, this values 4.9 eV). The recognized pattern of uniformly separated minima is proof that the atomic energy levels have to be discrete.

What about the impact of adjusting the temperature of the tube?

An increase in tube temperature would increase the stochastic thermal motion of the gas atoms inside the tube. In this way, the electrons have a significant probability of having more elastic collisions and getting a longer path to the anode. A longer route delays the time to touch the anode. Consequently, rising temperature increases the average time for the electrons to traverse the tube and reduces the current. The current declines as temperature rises, and the amplitude of the Franck-Hertz curves will fall, but the different patterns will remain.

CHAPTER 8:

ENTANGLEMENT

Quantum entanglement is considered one of science's most complicated concepts, but the fundamental problems are straightforward. Entanglement, once understood, allows for a deeper interpretation of ideas like quantum theory's "multiple worlds."

The concept of quantum entanglement, as well as the (somehow related) assertion that quantum theory involves "many worlds," exudes a glitzy mystique. Yet, in the end, those are scientific concepts with real consequences and definitions. Here, I'd like to describe entanglement and multiple worlds as quickly and explicitly as possible.

Entanglement is sometimes mistakenly thought to be a quantum-mechanical phenomenon, but it is not. In reality, it is instructive, although unorthodox, to first consider a non-quantum (or "classical") variant of entanglement. This concept allows one to separate the subtlety of entanglement from the general strangeness of quantum theory.

Entanglement happens when we have a partial understanding of the state of two systems. Our systems, for example, can be two objects we'll refer to as c-ons. The letter "c" stands for "classical," but if you want to think of something unique and fun, think of our c-ons as cakes.

Our c-ons are available in two shapes: square and circular, which we use to describe their possible states. For two c-ons, the four possible joint states are (square, square), (square, circle), and (circle, square) (circle, circle). The tables show two examples of the probabilities for finding the system in each of the four states.

INDEPENDENT

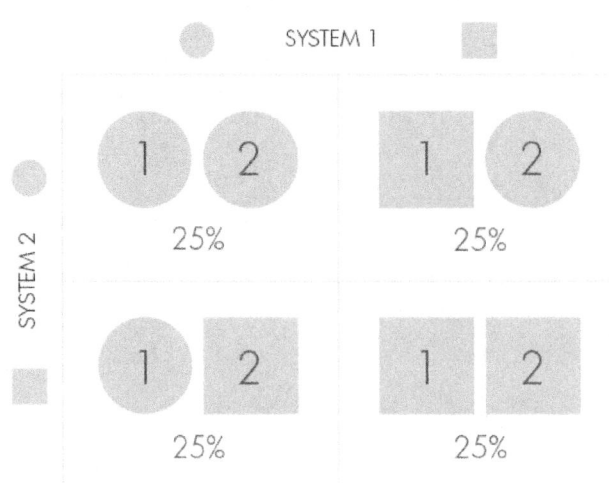

If knowing the state of one of the c-ons does not provide valuable information about the state of the other, we call them "independent." This state is a characteristic of our first table. If the form of the first c-on (or cake) is square, the second remains a mystery. Similarly, the state of the second reveals nothing informative about the first's shape.

When knowledge about one c-on enhances our understanding of the other, we say our two c-ons are entangled. The entanglement in our second table is extreme. When the first c-on is circular, we know the second would be circular as well. When the first c-on is square, the second is as well. Knowing the shape of one, we can easily predict the shape of the other

ENTANGLED

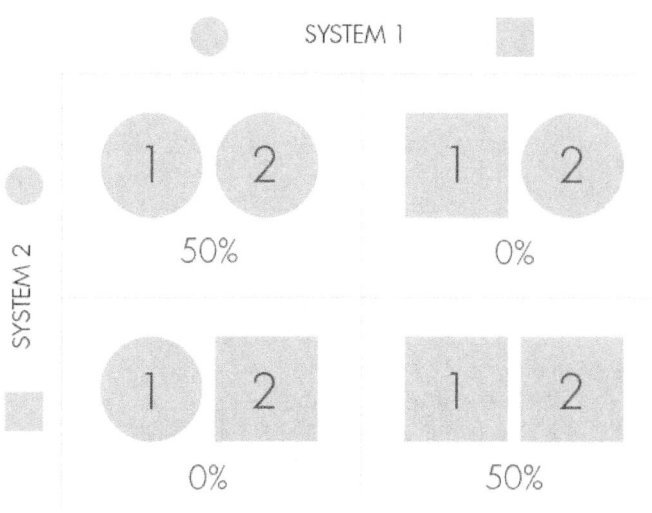

The quantum version of entanglement, or lack of independence, is the same as the previously explained phenomenon. Wave functions are mathematical objects that explain states in quantum theory. As we'll see, the rules linking wave functions to physical probabilities add some fascinating complications, but the core principle of entangled information, as we've seen before with classical probabilities, remains the same. Of course, cakes aren't quantum systems, but entanglement between quantum systems occurs naturally, such as in the aftermath of particle collisions. Unentangled (independent) states are unusual exceptions in practice since they establish correlations, take, for instance, molecules. They're made up of subsystems such as electrons and nuclei. The lowest energy state of a molecule, in which it is most commonly found, is a strongly entangled state of its electrons and nuclei since their locations are far from independent. The electrons follow the nuclei as they move.

Returning to our working example, if we have $\Phi_\blacksquare$, $\Phi_\bullet$ for the wave functions related to the system 1 in its square and circular states, and $\psi_\blacksquare$, $\psi_\bullet$ for the wave functions related to system 2 in its square and circular states, then the overall states will be in our working example:

Independent: $\Phi_\blacksquare \psi_\blacksquare + \Phi_\blacksquare \psi_\bullet + \Phi_\bullet \psi_\blacksquare + \Phi_\bullet \psi_\bullet$.

Entangled: $\Phi_\blacksquare \psi_\blacksquare + \Phi_\bullet \psi_\bullet$.

The independent version is also:

$(\Phi_\blacksquare + \Phi_\bullet)*(\psi_\blacksquare + \psi_\bullet)$

Note how the parentheses in this formulation explicitly divide systems 1 and 2 into separate units. Entangled states can be formed in several ways. One method is to measure your (composite) structure, which will provide you with only partial information. For example, we may discover that the two systems have conspired to have the same shape without learning what shape they have. Later on, this principle would be essential. The Einstein-Podolsky-Rosen (EPR) and Greenberger-Horne-Zeilinger (GHZ) results, for example, are the product of quantum entanglement's interaction with another aspect of quantum theory known as "complementarity." Let me now introduce complementarity as a prelude to discussing EPR and GHZ.

Previously, we thought our c-ons could take two different forms (square and circle). We can now picture it displaying two colors: red and blue. If we're talking about classic systems like cakes, this extra property means that our c-ons could be in one of four states: red square, red circle, blue square, or blue circle.

The situation is vastly different for a quantum cake, such as a quake or (with more dignity) a q-on. A q-on can show different shapes or colors in other circumstances, and this does not necessarily imply that it has both a shape and a color concurrently. As we'll see shortly, Einstein's "common sense" inference, which he believed should be part of every acceptable notion of physical truth, is inconsistent with experimental evidence.

We can calculate the shape of our q-on, but we will lose all its color information in the process. Alternatively, we can measure the color of our q-on, but we will lose all knowledge about its shape in the process. According to quantum theory, we can't calculate both the form and color at the same time. No single perspective on physical reality can cover all of its aspects; instead, several different, mutually exclusive views must be considered, each providing true but incomplete insight. As Niels Bohr put it, this is the heart of complementarity.

As a result, the quantum theory allows one to exercise caution when assigning physical reality to particular properties. To prevent inconsistency, we must agree that:
1. A property that cannot be calculated does not have to exist.
2. Measurement is a complex process that changes the system under analysis.

Quantum Physics for Beginners

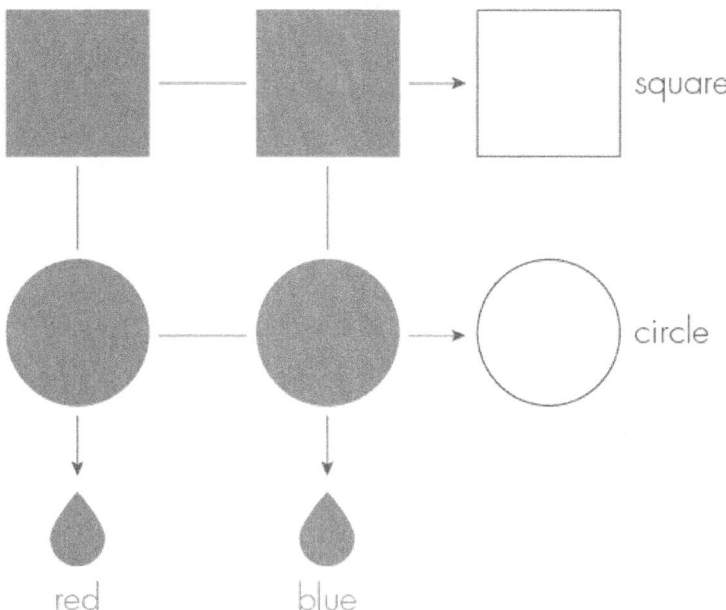

II.
Now we'll bring you two classic — but far from classic! — examples of quantum theory's strangeness. Both have been challenged in a battery of experiments. (In fact, people measure properties like electron angular momentum rather than the colors or shapes of cakes.)

When two quantum systems are entangled, Albert Einstein, Boris Podolsky, and Nathan Rosen (EPR) identified a surprising effect. The EPR effect combines complementarity with a particular, experimentally realizable type of quantum entanglement.

A pair of EPRs is made up of two q-ons, each of which can be measured for its shape or color (but not for both). We assume we have access to a large number of such pairs, all of which are similar, and that we can choose which of their components measurements to take. When one of the members of an EPR pair is measured, we find that it is equally likely to be square or circular, and the color is equally likely to be red or blue if we weigh it.

When we test both pair members, we get the curious results that EPR thought were paradoxical. We find that when we test all members for color or form, the results are still consistent. As a result, if we find one is red and then calculate the color of the other, we will find that it is red as well, and so on. On the other side, there is no connection between the shape of one and the color of the other. The second is equally likely to be blue or red if the first is square. Even if the two systems are separated by great distances and the measurements are performed virtually simultaneously, quantum theory predicts that we will get the results mentioned above. The measurement method used in one place seems to affect the condition of the system in the other. This "scary action at a distance," as

Einstein put it, seems to necessitate the transmission of data — in this case, data about the measurement — at a rate faster than the speed of light.

Does it, however, work? I don't perceive what to expect until I know what result you got. When I learn the outcome you've calculated, not when you figure it, I gain valuable knowledge. Any message disclosing the work you calculated must be sent in some physical form, probably slower than the speed of light.

The paradox becomes more evident as you think about it. Indeed, provided that the first system has been determined to be red, let us consider the condition of the second system once more. If we want to test the color of the second q-on, we will almost certainly get red. However, when implementing complementarity, if we're going to calculate a q-shape when it is "red," we will have an equal chance of finding a square or a circle. Rather than creating a paradox, the EPR result is logically induced. It's nothing more than a repackaging of complementarity.

It's also not paradoxical to discover that distant events are related. After all, if I put each member of a pair of gloves in a box and mail them to the opposite sides of the globe, I shouldn't be surprised if I can tell the glove's handedness in one box by looking inside the other. Similarly, the correlations between members of an EPR pair are imprinted in all known cases while they are close together. However, they may survive subsequent separation as if they had memories. The peculiarity of EPR is not so much the connection itself as it is the likelihood of its embodiment in complementary forms.

Daniel Greenberger, Michael Horne, and Anton Zeilinger discovered another beautifully illuminating example of quantum entanglement. It entails three of our questions, which have been prepared in a unique, entangled state (the GHZ state). Three distant experimenters are given the three q-ons. Each experimenter decides whether to test shape or color at random and records the results.

Each experimenter comes up with the most random results possible. When he measures the form of a q-on, he is equally likely to find a square or a circle; when he measures the color, he is equally likely to see red or blue. So far, everything has been so routinely.

However, when the experimenters compare their measurements later, a little study shows a surprising finding. Let's say square shapes and red colors are "good," while circular shapes and blue colors are "evil."

When two of the experimenters wanted to measure shape while the third measured color, they discovered that precisely 0 or 2 findings were "evil" (circular or blue); however, when all three decided to quantify color, they found that one or three of the measures were bad. Quantum mechanics predicts this, and it is confirmed by observation.

So, is the amount of evil even or odd? In different types of measurements, all possibilities are realized with certainty. We are compelled to say no to the issue. There is no point in talking about the quantity of evil in our system without considering how it's calculated. It does, in reality, lead to inconsistencies.

In physicist Sidney Coleman's words, the GHZ effect is "quantum mechanics in your face." It debunks a deeply embedded prejudice based on daily experience that physical systems have definite properties, regardless of whether those properties can be measured. And if they did, calculation options would not affect the balance between good and evil. The GHZ effect's message is unforgettable and mind-expanding until internalized.

We've looked at how entanglement makes it challenging to allocate individual, autonomous states to multiple q-ons so far. The evolution of a single q-on over time is subjected to similar considerations.

When it's hard to allocate an actual state to our system at any given time, we claim we have "entangled histories." We can establish entangled histories by making measurements that gather partial information about what happened, similar to how we got traditional entanglement by removing some possibilities. We only have one q-on in the most detailed entangled histories, which we monitor at two separate times. We can imagine scenarios in which we conclude that our q-on was either square or circular at both times, but our observations leave both possibilities open. The most uncomplicated entanglement conditions shown above have a quantum temporal equivalent. We may add the wrinkle of complementarity to this scheme and describe conditions that bring out the "many worlds" feature of quantum theory by using a slightly more elaborate protocol. As a result, our q-on may have been prepared in the red state at one point and then tested in the blue state at a later point. As in the single examples, we cannot consistently attribute our q-on the property of color at intermediate conditions, nor does it have a determinate shape. In a confined but measured and precise way, histories of this kind realize the intuition that bears the many-worlds picture of quantum mechanics. A solid state can branch into historical trajectories that are mutually contradictory but ultimately come together.

The evolution of quantum systems inevitably leads to states that can be calculated to have vastly different properties, according to Erwin Schrödinger, a pioneer of quantum theory who was highly suspicious of its correctness. His famed "Schrödinger cat" states scale-up quantum uncertainty into feline mortality issues. As we've seen in our examples, one can't assign the cat the property of life (or death) until it's measured. In a netherworld of possibility, both — or neither — coexist.

Quantum complementarity is challenging to explain in everyday language because it is not encountered in everyday life. Practical cats engage with surrounding air molecules in many different ways depending on whether they are alive or dead, so the measurement is made automatically in practice. The cat continues its life (or death). In a way, entangled histories characterize q-ons that are Schrödinger kittens. At intermediate times, we must consider both of two contradictory property-trajectories to describe them fully. Since it allows us to gather partial knowledge about our q-on, the controlled experimental realization of entangled histories is delicate. Rather than partial information spanning several times, conventional quantum measurements typically collect complete information at one time; for example, they decide on a definite shape or a definite color. But it can be done without much technical difficulty. This way, we can give the proliferation of "multiple worlds" in quantum theory a mathematical and experimental sense, demonstrating its substantiality.

Example

Assume that you and your friend are each given a small, opaque black box. Each box includes a standard six-sided die. To scramble the dice, you're both advised to shake your boxes lightly. After that, you split ways. Your companion returns to Australia while you return to the United States. During the process, you don't communicate with each other. You each open your box when you arrive home and examine the upward-facing number on your die.

Ordinarily, the numbers you and your friend see would have no association. Your friend is just as likely as you are to notice any number between 1 and 6; more importantly, the number they see on their die does not influence the number you see on yours.

Assume I now instruct your friend and you to lightly tap your boxes together before shaking them individually and going your separate ways. This isn't shocking; it's how the world works, after all. If we could make this example "quantum," though, it might act quite differently.

In a quantum mechanics example, tapping the boxes against each other would enchant the dice and link – or entangle – them in a mysterious way: once you each get home and open your boxes, you will discover that looking at the numbers, your number and your friend's number are sure to be precisely correlated. If you see a '4' in the United States, you can bet your buddy in Australia will measure a '4' on her die as well; if you see a '6,' they will as well.

The dice represent individual particles (such as atoms or light particles known as photons). The physical act of tapping the boxes together entangles them, allowing us to measure one die and learn about the other.

CHAPTER 9:

PARTICLES CREATION AND ANNIHILATION

In particle physics, annihilation happens when a subatomic particle collides with its antiparticle to create other particles, such as two photons, when an electron collides with a positron. The initial pair's total energy and momentum are conserved during the process and distributed among a group of other particles in the final state. Antiparticles have additive quantum numbers that are opposite those of particles, so the sums of all quantum numbers in such an original pair are zero. Consequently, any set of particles whose total quantum numbers are zero can be generated until energy and momentum conservation laws are followed.

Photon generation is preferred during a low-energy annihilation because photons have no mass. On the other hand, high-energy particle colliders produce annihilations, which produce a wide range of exotic heavy particles.

Informally, the term "annihilation" refers to the interaction of two but not reciprocal antiparticles (that is, they are not charged conjugate). Some quantum numbers do not amount to zero in the initial state, but they may conserve in the final state with identical sums. The "annihilation" of a high-energy electron antineutrino by an electron to generate a W- is an example. If the annihilating particles are composite, the final state usually produces several different particles. The opposite of annihilation is pair creation. Particles will mysteriously come out of nowhere. An energetic photon is used to create matter (gamma-ray). A photon is absorbed in the pair manufacturing process, resulting in a particle and its antiparticle, like an electron and positron pair. According to physics, energy is conserved since the photon's energy must be the same or greater than the sum of the energy of the two particles. However, the appearance of two particles at random is extremely strange. Only in the quantum universe does something like this happen. Two artifacts in the visible universe will never vanish and reappear at random.

We will now describe the most common annihilation examples.

Electron–positron annihilation

When a low-energy electron and a low-energy positron (antielectron) annihilate, the most likely outcome is the production of two or three photons. This production occurs because the only other final-state in the Standard Model is that electrons and positrons have enough mass-energy to create neutrinos, which are approximately 10,000 times less likely to produce. Momentum conservation prevents the creation of only one photon. The rest energy of the annihilating electron and positron particles is around 0.511 million electron-volts (MeV). This total rest energy is the photon energy of the photons emitted if their kinetic energies are negligible. The photons have an energy of around 0.511 MeV each.

Both momentum and energy are conserved, with 1.022 MeV of photon energy (which accounts for the particles' rest energy) traveling in opposite directions (accounting for the total zero momentum of the system).

Various other particles may be formed if one or both charged particles have higher kinetic energy. Furthermore, in the presence of a third charged particle, the annihilation (or decay) of an electron-positron pair into a single photon will occur, with the excess momentum transferred to the third charged particle by a virtual photon from the electron or positron. In a third particle electromagnetic field, the inverse process, pair generation by a single real photon, is also possible.

Proton–antiproton annihilation

When a proton collides with its antiparticle, the reaction is not as straightforward as electron-positron annihilation. A proton, unlike an electron, is a composite particle made up of three "valence quarks" and an unknown number of "sea quarks" bound together by gluons. To describe the interaction, we should go into some advanced particles physics description that is not in the scope of this book. If you want to think at a high level about what happens, you will consider that some proton and antiproton constituents will annihilate. Some others will rearrange in other particles that will not be stable. The result will be the emission of these unstable particles that will decay to stable particles emitting photons, plus other photons (also high energy ones). Antiprotons can annihilate with neutrons, and likewise, antineutrons can annihilate with protons. When an antinucleon annihilates within a more complex atomic nucleus, similar reactions will occur. The absorbed energy, which can be as high as 2 GeV, can theoretically surpass the binding energy of even the most massive nuclei. When an antiproton annihilates within a heavy nucleus like uranium or plutonium, the nucleus may be partially or fully broken, releasing many fast neutrons. These reactions can cause many secondary fission reactions in a subcritical mass, which could be helpful for spacecraft propulsion.

According to an experiment run at the CERN laboratory in Geneva Massive Hadron Collider in 2012 (LHC), the debris from proton-proton collisions resulted in the discovery of the Higgs Boson.

CHAPTER 10:

QUANTUM TUNNELING

What is the quantum tunneling effect? According to classical physics theories, it is a quantum phenomenon that occurs as particles pass through a barrier that should be difficult to pass through. A physically impenetrable medium, such as an insulator or a vacuum, or a region of high potential energy, may serve as a barrier. A particle with insufficient energy will not be able to overcome a potential obstacle in classical mechanics. However, in the quantum universe, particles often behave like waves. A quantum wave will not abruptly stop when it hits a barrier; instead, its amplitude will decrease exponentially. The probability of finding a particle in the deeper end of the barrier decreases as the amplitude decreases. If the barrier is thin enough, the amplitude on the other side can be non-zero. This will mean that any of the particles have a finite chance of tunneling through the barrier. The current density ratio from the barrier divided by the current density incident on the barrier is the tunneling current. There is a finite chance of a particle tunneling through the barrier if this propagation coefficient is non-zero.

Discovery of Quantum Tunneling

F. Hund discovered the possibility of tunneling in 1927. He measured the ground state energy in a "double-well" potential—a system in which a potential barrier separates two different states of equal energies. This form of mechanism can be found in many molecules, such as ammonia. Classical mechanics forbids "inversion" transformations between two geometric states, but quantum tunneling allows them. When observing the reflection of electrons from various surfaces in the same year, L. Nordheim discovered another tunneling phenomenon. Oppenheimer used tunneling to measure the ionization rate of hydrogen for the next few years. Garnow, Gurney, and Condon's study helped to clarify the spectrum of alpha decay rates of radioactive nuclei.

Quantum Tunneling in Nature

Tunneling, like quantum mechanics, can seem to have little significance to daily life, but it is an essential natural mechanism that is responsible for many things that life depends on. It has also been proposed that a tunneling event occurred at the very beginning of the universe, causing the universe to transition from a "state of no geometry" (no space or time) to a state where space, matter, time, and life could exist.

Tunneling in Stellar Fusion

Fusion is when small nuclei combine to form larger nuclei, releasing massive quantities of energy in the process. Except for hydrogen, fusion within stars produces all of the elements of the periodic table. The process of hydrogen fusion into helium is what gives stars their power.

Fusion happens much more often than previously believed. Since all nuclei are positively charged, they repel each other intensely, and their kinetic energy is rarely enough to overcome this repulsion and enable fusion to occur.

However, when tunneling effects are taken into account, the proportion of hydrogen nuclei capable of fusion rises significantly. This phenomenon explains how stars can sustain their stability over millions of years. However, science does not wholly endorse the procedure since an average hydrogen nucleus can encounter over 1000 head-on collisions before fusing with another.

Tunneling in Smell Receptors

Chemical receptors in the nose (400 different types in humans) were once thought to detect the presence of various chemicals via a lock-and-key mechanism that established the molecule's physical shape; however, more recent studies have found flaws in this hypothesis.

For example, despite their similar shapes, ethanol and ethanethiol have somewhat different smells (ethanol is the drinkable alcohol, while ethanethiol smells of rotten eggs).

This indicates the presence of another identifying mechanism. According to the theory, smell receptors use quantum tunneling to distinguish chemicals. The receptors send a small current through the odorant molecule, which causes it to vibrate in a specific pattern.

The electrons must tunnel into the non-conducting gap between the receptor cells and the molecule for the current to flow.

Large hydrogen and deuterium bonds were used in new quantum tunneling experiments to promote olfactory reactions to stimuli. The findings revealed higher quantum vibrations, implying that humans can distinguish between molecules with quantum vibrational signatures.

Applications of Quantum Tunneling

Josephson Junctions

A very thin layer of non-superconducting material separates two superconductors, and this may be an insulator, a non-superconducting metal, or a physical defect in Josephson junctions. The superconducting current will tunnel through the barrier since the system's electrical properties are well defined. This opens up new possibilities, most notably for ultra-precise measurements. Josephson junctions are commonly used in superconducting electronics, quantum computers, and superconducting quantum interference devices (SQUIDs), which can detect extremely weak magnetic fields. The use of these junctions to calculate quantum coherence is also being investigated.

Tunnel Diodes

A tunnel diode is a high-speed semiconductor that consists of a thin insulator sandwiched between two semiconductors. This diode, also known as the Esaki diode after L. Esaki's work on the topic, can operate at frequencies higher than the microwave range.

Quantum tunneling allows the electric current to pass through tunnel diodes and two-terminal devices with a highly doped p-n junction. As the voltage rises, the current which passes through the tunnel diode decreases significantly. As a result, the tunnel diode's resistance would be negative. A tunnel diode has 1000 times more impurity concentration than a standard p-n junction diode. As a result, the depletion area of the p-n junction is tiny on the order of nanometers. When you apply a higher voltage than the depletion region's built-in voltage, the current is generated in ordinary diodes.

Due to quantum tunneling among the n and p regions, a small voltage less than the depletion field's built-in voltage induces an electric current in tunnel diodes. The narrow depletion region is needed for tunneling because the barrier thickness must be kept low.

Tunnel diodes are used as ultra-high-speed switches, relaxation oscillator circuits, and logic memory storage units. They are widely used in the nuclear industry due to their high radiation tolerance.

Scanning Tunneling Microscopes

A Scanning Tunneling Microscope (STM) scans the surface of a substance with a very sharp conducting probe. An electrical current is sent down the probe's tip, tunneling through the gap and into the material.

The tunneling current gets tinier or more significant as the distance gets broader or narrower, respectively. A highly detailed image of the surface could be produced, and it can even resolve surface humps caused by individual atoms. The physics and chemistry of surfaces have been better understood thanks to this methodology.

Flash Drives

Tunneling is one of the causes of current leakage in very large-scale integration (VLSI) electronics, which results in significant power drain and heating effects. It is thought to be the lowest point at which microelectronic system elements can be manufactured. Tunneling is an essential technique for programming flash memory's floating gates. On flash drives, data is stored in a network of memory cells made up of "floating-gate" transistors. Two metal gates, a control gate, a floating gate make up this collection. The floating gate is encased in a metal oxide insulating film. In its natural state, a floating-gate transistor registers a "1" in binary code. When an electron attaches to the floating gate, it gets entangled in the oxide layer, causing the control gate voltage to shift. In this condition, a transistor registers a binary "0."

As data from flash memory is erased, a strong positive charge added to the control gate allows the trapped electron to tunnel through the insulating layer, returning the memory cell to a "1" state.

Cold emission
The physics of semiconductors and superconductors is affected by cold electron emission. It's analogous to thermionic emission. Electrons leap from a metal's surface to follow a voltage bias because they statistically have more energy than the barrier due to random collisions with other particles. When the electric field is sufficiently high, the barrier becomes thin enough for electrons to tunnel out of the atomic state, resulting in a current that varies exponentially with the electric field. Flash memory, vacuum tubes, and specific electron microscopes all need these materials.

Quantum-dot cellular automata
QCA, a molecular binary logic synthesis technology, operates by the inter-island electron tunneling device powers. This is a very low-power, high-speed device with a maximum frequency of 15 PHz.

Tunnel field-effect transistors
Field effect transistors with the gate (channel) operated by quantum tunneling rather than thermal injection were demonstrated in a European research project, and this exhibited a decrease of gate voltage from 1 volt to 0.2 volts and a decrease in power consumption by up to 100%. If these transistors can be improved into VLSI chips, integrated circuit output per power would improve.

Nuclear fusion
Nuclear fusion depends on the phenomenon of quantum tunneling. The temperature in the nuclei of stars usually is too low for atomic nuclei to break through the Coulomb barrier and achieve thermonuclear fusion. Quantum tunneling increases the chances of breaking through this barrier. The massive number of nuclei in a star's center is sufficient to support a steady fusion reaction–a precondition for developing life in insolation habitable zones though the likelihood of this is still lacking.

Radioactive decay
The process of emitting particles and energy from an atom's unstable nucleus to form a stable substance is known as radioactive decay. A particle tunneling accomplishes this out of the nucleus (an electron tunneling into the nucleus is electron capture). This was the first time someone used quantum tunneling. In conditions beyond the circumstellar habitable zone where insolation will not be feasible (subsurface oceans), radioactive decay is a vital problem for astrobiology since it produces a continuous energy source over a long period as a result of quantum tunneling.

Astrochemistry in interstellar clouds
The astrochemical syntheses of different molecules in interstellar clouds, such as the synthesis of molecular hydrogen, water (ice), and the prebiotically essential formaldehyde, can be explained by including quantum tunneling.

Quantum biology

In quantum biology, quantum tunneling is one of the essential non-trivial quantum effects. Both electron and proton tunneling are crucial in this case. Several biochemical redox reactions (cellular respiration, photosynthesis) and enzymatic catalysis depend on electron tunneling. In spontaneous DNA mutation, proton tunneling is a crucial factor.

When regular DNA replication occurs after a particularly large proton has tunneled, a spontaneous mutation occurs. A hydrogen bond joins the base pairs of DNAS. A potential energy barrier is isolated by a double-well potential along with a hydrogen bond. The double-well potential is thought to be asymmetric, with one well deeper than the other and the proton usually resting in the deeper well. The proton would have tunneled through the shallower well for a mutation to occur. A tautomeric transition is the displacement of a proton from its normal position. If DNA replication occurs in this state, the DNA base pairing rule can be compromised, resulting in a mutation. Per-Olov Lowdin first proposed this theory of spontaneous mutation inside the double helix. Other quantum tunneling-induced biological mutations are thought to be a cause of aging and cancer.

Quantum conductivity

Although the Drude model of electrical conductivity gives good predictions about the nature of electrons conducting in metals, quantum tunneling can describe the nature of electron collisions. When a free electron wave packet comes into contact with a long array of uniformly spaced barriers, the wave packet's reflected portion interferes uniformly with the transmitted one between all barriers, allowing for 100% transmission. According to the theory, if positively charged nuclei form a perfectly rectangular series, electrons will tunnel as free electrons through the metal, resulting in remarkably high conductance. However, impurities in the metal will significantly disrupt it.

Scanning tunneling microscope

Heinrich Rohrer and Gerd Binnig developed the scanning tunneling microscope (STM), which enables imaging of individual atoms on a material's surface. It functions by exploiting the relationship between quantum tunneling and space. When the STM's needle is brought close to a voltage-biased conduction surface, the current of electrons tunneling between the needle and the surface is measured, revealing the distance between the needle and the surface. The height of the tip can be changed to keep the tunneling current constant by using piezoelectric rods that change size as the voltage is applied. These rods may have time-varying voltages applied to them, which can be registered and used to picture the conductor's surface. STMs have a resolution of 0.001 nm or around 1% of the atomic diameter.

Kinetic isotope effect

In chemical kinetics, replacing a light isotope of an element with a heavier one causes the reaction rate to slow down. This is due to variations in the zero-point vibrational energies for chemical bonds comprising of lighter and heavier isotopes and is modeled using transition state theory. However, in some cases, significant isotope effects are found and cannot be explained by a semi-classical treatment, necessitating the use of quantum tunneling. R. P. Bell developed a modified Arrhenius kinetics treatment that is widely used to model this phenomenon.

CHAPTER 11:

THE COMPTON EFFECT

A photon interaction with an electron is known as the Compton effect by physicists. The photon loses energy after bouncing off a target electron. As gamma moves through matter, these elastic collisions compete with the photoelectric effect, leading to the weakening of its signal. Arthur H. Compton, an American physicist, discovered the effect in 1922. In 1927, the Nobel Prize in Physics was awarded to Compton. He showed electromagnetic radiation's particle existence. At the time, it was a significant development. As a graduate of the University of Wooster and Princeton University, Compton developed a theory regarding the strength of X-ray reflection from crystals to observe the arrangement of electrons and atoms. He began researching X-ray scattering in 1918. Compton received one of the first National Research Council fellowships in 1919. He took his organization to the Cavendish Laboratory in Cambridge, England. When the facilities in England proved insufficient for his needs, he then moved to Washington University in St. Louis. He perfected his X-ray apparatus used to calculate the wavelength shift with scattering angle, known as the Compton effect.

Compton measured the wavelength of X-rays scattered by electrons in a carbon target and discovered that the scattered X-rays were longer than those incidents on the target. With the scattering angle increase, the wavelength shift increased. Compton clarified and modeled the data by assuming that light is a particle (photon) and applies the conservation of momentum and energy to the collision of the photon and the electron. According to the relationship between wavelength and energy discovered in 1900 by German physicist Max Planck, the scattered photon wavelength is longer and the energy lower.

When X-ray or gamma-ray photon interacts with matter, the Compton effect occurs, and it is characterized as a decrease in energy (increase in wavelength). This effect shows that light needs to be explained, not solely in terms of waves. Compton's work showed that light acts as a stream of particles whose energy is proportional to frequency in scattering experiments (i.e., inversely proportional to the wavelength).

The electron receives a portion of the energy due to the interaction, and a photon containing the remaining energy is released in a different direction than the initial, preserving the system's overall momentum.

Compton collisions can be seen as elastic collisions between an electron and a photon. When the photon energy exceeds the energy that maintains the electron in an atom (the binding energy), elastic collisions become prevalent. The Compton effect dominates the photoelectric effect when a light atom like carbon goes over 20 keV. It is larger than 130 keV for copper and 600 keV for lead.

The phenomenon, which is very broad, affects all of the atom's electrons in this gamma energy range. In the photoelectric effect, only the two electrons from the innermost K shell play a role. In the area where the Compton effect dominates, the density of electrons is essential for an absorber. Lead thus has an advantage over lighter materials, even if it is less significant than the photoelectric effect, which was caused by the high electrical charge of its nucleus at the fourth power.

The collision does not harm the gamma. The "scattered" photon, which emerges with the electron, shares the initial energy with the electron set in motion. The electron then ionizes and loses its energy as a beta electron. Before it interacts again, the scattered gamma propagates through the material without depositing energy.

The distribution of energy is unequal. The scattered photon and the original gamma's angle determine this (the gamma possibility distribution at a provided angle is produced by a formula called the "Klein-Nishima formula"). Given its shallow mass, the electron is a formidable target for a massless photon. The scattered photon carries the majority of the initial energy, on average of 96% at 50 keV and 83% at 500 keV, according to the laws of physics that regulate the Compton effect. The emerging direction of the scattered photon is usually the same as the incident photon. There is also the case of backscattering that occurs when it goes backward. On average, it scatters at a 30-to-45-degree angle. Gamma of hundreds of keV will undergo multiple Compton scattering before being absorbed by the photoelectric effect.

A new phenomenon threatens Compton scattering: transforming a gamma photon into an electron and its antiparticle, a positron when the gamma energy reaches 1 MeV, which is unusual for gamma rays emitted by nuclei. This phenomenon is mainly produced when high-energy gamma is produced, as in particle accelerators.

Since it is the most likely association of high-energy X-rays with atomic nuclei in living organisms and is used in radiation therapy, Compton scattering is of critical interest to radiobiology. Compton scattering can be used to investigate the wave function of electrons in matter in material physics.

As shown before, Compton also discovered the phenomena of total X-ray reflection and full polarization, which allowed for a more precise calculation of an atom's number of electrons. He was also the first to obtain X-ray spectra from ruled gratings, which allows for direct measurement of X-ray wavelength. It is possible to calculate the absolute value of the spacing of atoms in the crystal lattice by comparing these spectra to those obtained using a crystal. As a result, a new electronic charge measurement was created.

CHAPTER 12:

WAVE-PARTICLE DUALITY

An electromagnetic wave carries the energy of the radiation detected by a radio signal transmitting antenna. The energy of individual photon particles is the same as the radiation observed by a photocurrent in the photoelectric effect. As a result, the nature of electromagnetic radiation arises: Is a photon called a wave or a particle? We can apply similar inquiries to other known sources of energy. An electron that is part of an electric current in a circuit, for example, behaves like a particle traveling in lockstep with other electrons inside the conductor. When an electron passes through a solid crystalline structure and forms a diffraction image, it behaves like a wave. Is an electron a wave or a particle? The same question may be applied to all particles of matter, both elementary and compound molecules, to determine their true physical existence. Such questions about the fundamental nature of things do not have definitive answers in our current state of understanding. All we can conclude is that wave-particle duality occurs in nature: a particle appears to behave as a particle under some experimental conditions, and a particle seems to act as a wave under different experimental conditions. In certain physical situations, electromagnetic radiation behaves like a wave, and in others, it behaves like a ray of photons. This dualistic interpretation is not a brand-new physics phenomenon spawned by specific twentieth-century discoveries. It was already present in a debate about the existence of light between Isaac Newton and Christiaan Huygens that began in 1670. A ray of light, according to Newton, is a series of light corpuscles. Light, according to Huygens, is a wave. In 1803, when Thomas Young published his double-slit interference experiment with light (see figure below), showing that light is a wave. He debunked the corpuscular hypothesis according to James Clerk Maxwell's electromagnetism theory (completed in 1873).

While Maxwell's classical view of radiation as an electromagnetic wave is still valid today, it fails to account for blackbody radiation and the photoelectric effect, in which light acts as a beam of photons.

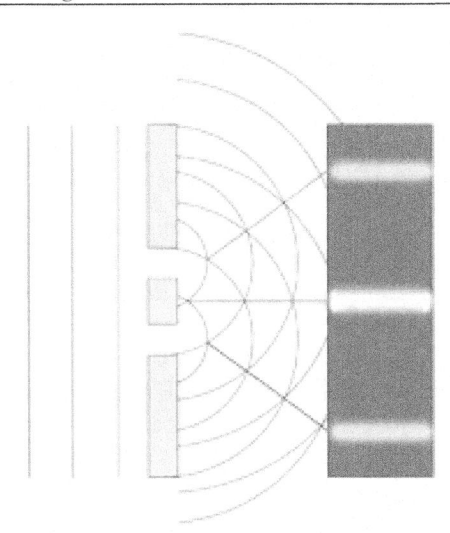

Young's double-slit experiment compares light interference to the interference of water waves and describes it. At the locations of two slits in an opaque panel, two waves are produced. The wavelengths of the waves are identical. They pass from the slits to the display screen, which is situated to the right of the slits. On the display screen, the waves collide. The meeting waves are in phase, and the combined wave amplitude is increased at the positions labeled "Max" on the screen. The cumulative wave amplitude is zero at the "Min" positions. This process produces a bright-and-dark fringe pattern on the viewing screen in the case of illumination.

In the understanding of electricity, there was a similar difference. Electric current was perceived as a flow in a continuous electric medium from Benjamin Franklin's electricity observations in 1751 until J.J. Thomson discovered the electron in 1897. This electric fluid theory helped develop the current theory of electric circuits and discover electromagnetism and electromagnetic induction. Thomson's experiment demonstrated that the unit of negative electric charge (an electron) would move in a vacuum without using a medium, such as in electric circuits. This discovery revolutionized our understanding of electricity and elevated the electron to the level of an atom. The electron and proton are also called matter particles in Bohr's early quantum theory of the hydrogen atom. Similarly, the electron is a particle in Compton scattering of X-rays on electrons. On the other hand, the electron acts like a wave in electron scattering experiments on crystalline structures.

A skeptic might argue that an electron is always just a photon. The diffraction images obtained in electron-scattering experiments can be explained using a macroscopic model of a crystal and a macroscopic model of electrons raining down on it like ping-pong balls. We don't need a complex crystal model to investigate this question; all we need are a few simple slits on a screen that is opaque to electrons. In other words, we need to replicate the Young double-slit experiment with electrons to gather compelling evidence about the existence of an electron.

Even when electrons pass through the slits one by one, if the electron is a wave, we can see interference patterns typical to waves, such as those shown in the figure below. The interference fringes will not form if the electron is a particle rather than a wave.

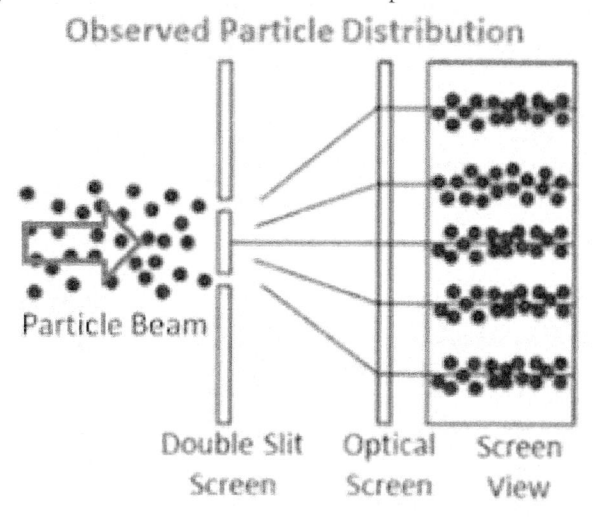

Observed Particle Distribution

Particle Beam

Double Slit Screen   Optical Screen   Screen View

Claus Jönsson's 1961 double-slit experiment with a beam of electrons showed that a beam of electrons does form an interference pattern, which indicates that electrons collectively behave as a wave. Giulio Pozzi in Italy and Akira Tonomura in Japan conducted the first double-slit experiments with single electrons crossing the slits one by one in 1974 and 1989, respectively. Even when electrons go through the slits individually, they show that interference fringes form gradually.

This experiment proves, beyond any doubt, that electron-diffraction images are created due to the wave existence in electrons. The pictures of the interference pattern in double-slit experiments with electrons demonstrate the findings shown in those experiments (Figure). In the Young double-slit experiment with electrons, it is possible to see computer-simulated interference fringes. Regardless of whether the electrons pass through the slits in a beam or one by one, a pattern gradually forms on the screen.

CHAPTER 13:

SCHRODINGER EQUATION

At the turn of the twentieth century, different experiments suggested that atomic particles were often wavelike. Electrons, for example, were discovered to produce diffraction patterns when passing through a double-slit in the same way as light waves do. As a result, it was rational to conclude that a wave equation could describe atomic particle behavior. Schrodinger was the first to write down a wave equation like this. The meaning of the equation sparked a lot of debate. The wave equation's eigenvalues were shown to be equal to the quantum mechanical system's energy levels. The equation's most significant test was used to solve the hydrogen atom's energy levels (found to be following Rydberg's Rule). It took time to figure out what the equation wave function was. The wave function is now accepted as a probability distribution after much discussion. The Schrodinger equation is used to calculate the quantum mechanical systems' permitted energy levels (such as atoms or transistors). The related wave function indicates the likelihood of finding the particle at a given location. The Schrodinger equation is the basic non-relativistic wave equation used to explain a particle's behavior in a field of force in one version of quantum mechanics.

There is a time-dependent equation for explaining progressive waves that can be applied to free particle motion. This equation's time-independent form is used to describe standing waves.

The Schrodinger equation:

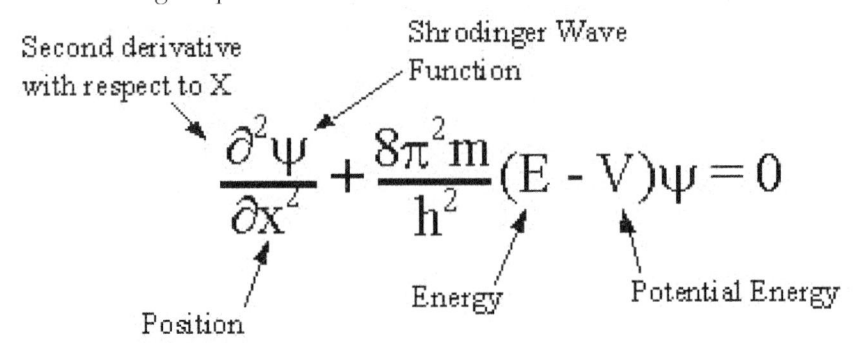

$$\frac{\partial^2 \psi}{\partial x^2} + \frac{8\pi^2 m}{h^2}(E - V)\psi = 0$$

- Second derivative with respect to X: $\frac{\partial^2 \psi}{\partial x^2}$
- Shrodinger Wave Function: ψ
- Position: x
- Energy: E
- Potential Energy: V

The solution to the equation is a wave that defines a system's quantum aspects. On the other hand, physically interpreting the wave is one of quantum mechanics' major philosophical issues.

The system of Eigen Values, devised by Fourier, is used to solve the equation. Any mathematical function can be expressed as the sum of an infinite number of other periodic functions in this way. The trick is to find the proper functions with the correct amplitudes so that they achieve the desired effect when they're superimposed.

As a result, the wave function for the structure, the solution to Schrodinger's equation, was replaced by the individual series' wave functions, natural harmonics of each other, and an infinite series. The replacement waves represented the individual states of the quantum system, and their amplitudes gave the relative significance of each state to the whole system, according to Schrödinger. Schrodinger's equation, which describes all of matter's wavelike properties, is considered one of science's most remarkable accomplishments of the twentieth century.

It is used to solve problems concerning the atomic structure of matter in physics and much of chemistry. It is a compelling mathematical method that forms the base of wave mechanics.

For a variety of simple systems, Schrodinger's time-independent equation can be solved analytically. Since the time-dependent equation is of the first order in time but of the second order in terms of coordinates, it is incompatible with relativity. For bound systems, the solutions give three quantum numbers, which correspond to three coordinates, and adding a fourth spin quantum number allows for an approximate relativistic correction.

CHAPTER 14:

QUANTUM FIELD THEORY

All particles are depicted as "excitations" that form in underlying fields in quantum field theory. It combines the notions of other quantum theories. With his equation defining how relativistic electrons – and thus most other matter particles – behave, British scientist Paul Dirac set the ball moving in the late 1920s.

Standard quantum theory, as developed in the 1920s by Niels Bohr and Werner Heisenberg, is adequate for describing the behavior of single particles in isolation and at moderate speeds. But you'll need something more to explain their interactions in the real world.

You'll need to combine quantum theory with special relativity, Einstein's theory of how space and time bend for fast-moving objects. As expressed by the equation $E=mc^2$, special relativity states that mass and energy are interchangeable. Meanwhile, according to Heisenberg's quantum uncertainty principle, particles can borrow energy from the vacuum for a limited duration.

The Dirac equation contained a catch: it predicted the creation of a particle that was identical to the electron in all respects except for the electric charge. The positron, the first antimatter particle, was identified in cosmic rays a few years later. It was the first of a slew of new particles postulated by theorists as quantum field theories advanced — and eventually discovered in reality.

The particle physics standard model is based on two quantum field theories. This model includes the workings of three of the four forces of nature through interactions of force-carrying boson particles with matter-making fermions. Experiments have rigorously confirmed the result of many years of theoretical effort.

Quantum electrodynamics (QED) is a unified "electroweak" theory of electromagnetism and the weak nuclear force that drives nuclear processes like radioactive beta decays, which are essential in how the sun consumes its fuel, for example.

In the meantime, quantum chromodynamics (QCD) is the theory of the strong nuclear force. This strong, very short-range force emitted by bosons called gluons binds quarks together to form particles like protons and neutrons.

The mass of a fundamental particle is dictated by its degree of interaction with the Higgs boson, which is the most solid feature of matter. The molasses-like field associated with the Higgs provides a drag that varies depending on particle type, according to a theory developed in 1964. The finding of the Higgs boson, predicted over five decades before, gave the standard model its crowning glory in 2012.

A quantum field theory of gravity, on the other hand, is still missing. Gravity, the only force without particles, is explained by Einstein's general theory of relativity as the warping of space-time, a whole different kettle of fish.

CHAPTER 15:

PERIODIC TABLE OF ELEMENTS

Dmitri Ivanovich Mendeleev suggested the periodic system 150 years ago. A fundamental chemistry law that was then clarified in terms of quantum physics. D. I. Mendeleev, then 34 years old, proposed his first version of what became identified as the periodic system or table to the newly founded Russian Chemical Society in the spring of 1869.

Mendeleev's method of chemical element classification established a firm basis for inorganic chemistry, bringing order to the confounding amount of experimental data while also predicting the presence of some previously unknown elements. In a nutshell, it was highly fruitful.

Even though today's periodic tables have 118 elements rather than the 63 in Mendeleev's original table, the systems in use today are direct descendants of the one proposed in 1869. The periodic system belonged solely to the study of chemistry until the end of the nineteenth century, but physicists took the lead in explaining why the system works so well. As discovered in the 1920s, the periodic system is a macroscopic representation of the internal structure of atoms. It's as much about physics as it is about chemistry from this perspective. Of course, knowing what makes up a chemical element is essential for the periodic system to function. The modern definition of an element dates back to the late eighteenth century and owes much to the great reformer of chemistry, Antoine-Laurent Lavoisier of France. Lavoisier's concept of an element was empirical because he restricted elements to substances that could not be decomposed into even simpler substances. Although this description is independent of whether or not matter is made up of atoms, in 1808, John Dalton made the crucial step of connecting elements and atoms. According to Dalton, the smallest unit of an element is the atom, and there are as many diverse atoms as there are different elements. Furthermore, Dalton developed the crucial concept of atomic weight, which linked an element to a measurable quantity. As a result, a pure element can be described as a substance with a specific atomic weight. The determination of atomic weights had become an integral part of chemistry by the late 1850s, when Dalton's theories had gained widespread acceptance. Mendeleev built his system of the elements on this foundation, and Julius Lothar Meyer, a German chemist, proposed a roughly similar scheme later in 1869. Mendeleev came first, and his method was more popular than Meyer's, so the periodic system has two fathers.

The periodic system was recognized as an important key to understanding the elements and their combinations into chemical compounds by the mid-1880s. There were essentially two big reasons why chemists were so excited about Mendeleev's invention; for the first time, there was an arrangement of all known elements in a single coherent structure, creating an organizing concept that applied to all elements and their properties. Second, and even more importantly, Mendeleev's method was highly predictive. It predicted the presence of new elements that were later discovered and modified the atomic weights of certain elements. For example, the beryllium atomic weight was once thought to be around 14.5, making it a homolog of aluminum, but Mendeleev claimed that his method wanted the atomic weight to be about 9. When experiments from around 1880 revealed that beryllium's atomic weight was 9.1, it was widely regarded as a victory for Mendeleev's classification scheme.

Mendeleev famously believed that the holes in his system corresponded to three unknown elements with atomic numbers of about 44, 68, and 72, similar to boron, aluminum, and silicon. The three elements were called scandium (1879), gallium (1875), and germanium (1886) when they were discovered in nature between 1875 and 1886, and their properties proved to be very similar to those predicted by Mendeleev.

For most chemists, these and other accurate predictions proved the periodic system's accuracy without a shadow of a doubt. Several of Mendeleev's predictions, such as his prediction of "Eka-caesium" with an atomic weight of about 175 and his later claim that the physicists' ether was a chemical element much lighter than hydrogen, were inaccurate. Despite its numerous successes, by the late nineteenth century, the periodic system had problems and inconsistencies, which exposed flaws in the original table. For example, Mendeleev's table implied that iodine must have a higher atomic weight than tellurium, but experiments revealed the opposite. When argon was found accidentally in 1894 and discovered to be a monoatomic gas with an atomic weight of 39.8, it caused problems. The new element seemed to have no place in the periodic system, leading Mendeleev to question its presence and suspect that it could be triatomic nitrogen N3 rather than an element. It was discovered that the danger of the periodic system only seemed to emerge after the discovery of helium and other noble gases. The original scheme required the addition of a new category for noble gases.

Nonetheless, the fact that argon's atomic weight (40.1) was between that of potassium (39.1) and calcium (39.2) posed a concern. Is it true that the atomic weight wasn't the proper ordering parameter for the periodic system after all?

According to Mendeleev and most chemists, the periodic system of the elements was first and foremost an empirical law that did not need further clarification in terms of, say, atomic structure. In reality, Mendeleev flatly refused to believe that atoms could be made up of smaller particles. However, a few speculative chemists were perplexed as to why certain atomic weights in the system had those values while others were lacking.

Other chemists hypothesized that all atoms were conglomerates of microscopic primordial particles, like hydrogen atoms or anything smaller, though Mendeleev dismissed such questions as nonsense. William Crookes of London, Julius J. Thomsen of Copenhagen, and Lothar Meyer of Tübingen, Germany, were among those who believed that the periodic mechanism could be described in terms of atomic structure. The majority of chemists, however, were not pleased with their vague and qualitative recommendations. They were turned into a quantitative theory when J. J. Thomson, a physicist rather than a chemist, discovered the electron at the turn of the century.

According to Thomson's 1904 atomic model, the atoms were made up of a large number of negatively charged electrons traveling in circular orbits within a positively charged weightless sphere. Thomson used this image to conduct complicated calculations to find stable electron configurations that he thought were similar to real atoms. In this way, he created configurations with a noticeable periodicity, similar to Mendeleev's table. Thomson's theory offered "an explanation for the empirical relations between atomic weight and atomic properties embodied in the periodic law," as chemist Ida Freund optimistically wrote in 1904.The chemical properties of the elements, according to Thomson's model, were correlated with specific electron structures that corresponded to and clarified the chemical groups. He did, however, align the groups with internal electron structures rather than the electrons in the outermost ring, as later theories did. Thomson's valiant attempt to describe the periodic system in terms of electron configurations was not only shaky, but it was also based on an erroneous atomic model that only lasted a decade. Nonetheless, it was the first attempt of its kind, and for that reason alone, it merits a place in science history.

Radioactivity discovery in 1896 and the realization that radioactive decay meant the instability of some elements posed challenges to the traditional understanding of elements and classification. According to Mendeleev, radioactivity was not an atomic property, and radium and other heavy elements did not naturally transmute into other elements. He claimed this was incompatible with the periodic law's fundamental tenets about the existence of elements.

A similar concern was the large number of radioactive compounds isolated from the uranium, thorium, and actinium decay sequence. Some of these compounds were chemically indistinguishable from other elements, but they were distinct from them and lacked periodic table positions.

In 1910, Frederic Soddy proposed that the same element could be divided into different species based on their atomic weights. The proposal was revolutionary because it contradicted the chemical theory, which stated that all parts of an element have the same atomic weight. It was "in complete contradiction to the theory of the periodic rule," as Soddy acknowledged. Three years later, he invented the term "isotope," which he now associates with Ernest Rutherford's atomic nucleus theory.

Many physicists had converted by 1915 to Rutherford and Niels Bohr's new image of the atom, which included the idea that the atomic number, as determined by the nuclear charge, was an element-defining parameter. Since two isotopes have the property of sharing the same atomic number but different atomic weights, they are in the same place in the periodic table and have the same chemical properties. Henry Moseley's pioneering X-ray spectroscopic measurements offered conclusive evidence of the atomic number principle. The Rutherford-Bohr model of the atom received good support as a result of this. Moseley's approach was crucial because it contributed to a new understanding of Mendeleev's system's essence.

Bohr proposed electron configurations for the light elements that roughly corresponded to the periodic system as early as 1913 in his atomic theory. Still, it wasn't until his later and more advanced theory that he established a full-scale explanation based on the old quantum theory's concepts. He described electron orbits by their principal and azimuthal quantum numbers, which were symbolized as n and k at the time, in works published between 1921 and 1923. He allocated electron configurations to all elements from hydrogen to uranium for the first time. This was driven by the correspondence theory, data from X-ray spectroscopy, and chemical properties of the elements. He was able to recreate the periodic system in this way, including the poorly known group of rare-earth metals, which Bohr believed included precisely 14 elements.

Bohr went much further in his Nobel Prize lecture of 1922, suggesting an electron configuration for the hypothetical element with atomic number 118. This element (oganesson) was recently synthesized in nuclear reactions, and its electron structure is assumed to be identical to that predicted by Niels Bohr nearly a century ago. Richard Swinne, a German physicist, speculated on transuranic elements when he assigned electron structures to elements with atomic numbers ranging from 92 to 105 in 1925. Bohr's reconstruction of the periodic system sparked widespread enthusiasm, although among atomic physicists rather than chemists. In the case of the enigmatic element 72, Bohr deduced that it had to be chemically similar to zirconium and not a rare planet, as most chemists assumed. In Copenhagen, two scientists at Bohr's institute, George von Hevesy and Dirk Coster, were inspired by Bohr's prediction and succeeded in detecting the element's characteristic X-ray lines in zirconium minerals. Hafnium discovery was widely regarded as a vindication of Bohr's theory as the new element was given. However, it was discovered that the impressive explanation was just a temporary solution to the puzzle of explaining the periodic system. Although the modern understanding of the periodic system is based on Werner Heisenberg and Erwin Schrödinger's quantum mechanics theory, it dates slightly earlier. Edmund Stoner, a British physicist, classified the energy levels in an atom by three quantum numbers rather than two in a critical revision of Bohr's theory in 1924. The result was a more complete and fine-grained system of electron configurations than Bohr's.

Stoner's theory inspired Wolfgang Pauli to publish a now-classic paper in Zeitschrift für Physik in March 1925, in which he presented the famous exclusion principle. Pauli was awarded the Nobel Prize for discovering "a new law of nature, the exclusion principle or Pauli principle," which he won belatedly in 1945.

In its most general form, the exclusion principle notes that two identical fermions cannot occupy the same quantum state. According to the theory applied to atomic electrons, it is unlikely for two electrons in an atom to have the same values as the four quantum numbers. The spin quantum number is one of the quantum numbers, with only two possible values: "spin up" and "spin down." The spin, like quantum mechanics, had not yet been discovered in March 1925. Instead, Pauli introduced the idea of an electron's "twofoldness" (Zweideutigkeit), which can be thought of as a precursor to spin.

On this basis, Pauli explained the details of the periodic system. Still, he also discovered that $N = 2n^2$ gives the maximum number N of elements in a time, where n denotes the principal quantum number n = 1, 2, ... In honor of Swedish physicist Janne Rydberg, the law is known as Rydberg's rule. Julius Thomsen proposed a similar rule in 1895, but it was when the noble gases were still a mystery.

CHAPTER 16:

LASERS

Laser means Light amplification by the stimulated emission of radiation. A laser is a system that causes atoms or molecules to emit light at specific wavelengths and amplifies it, resulting in a very narrow radiation beam. The emission is usually limited to visible, infrared, or ultraviolet wavelengths. There have been several different types of lasers produced, each with its own set of characteristics.

History

The laser is a result of Albert Einstein's 1916 notion that, under the right conditions, atoms may emit excess energy as light, either naturally or when stimulated by light. In 1928, German physicist Rudolf Walther Ladenburg detected stimulated emission for the first time, but it seemed to have little practical application at the time.

In 1951, Charles H. Townes, then a graduate student at Columbia University in New York City, invented a process for generating microwave-frequency stimulated emission. He demonstrated a working system at the end of 1953 that concentrated "excited" ammonia molecules in a resonant microwave cavity. They emitted a pure microwave frequency. The system was given the name maser by Townes, which stands for "microwave amplification by stimulated emission of radiation."

Theodore H. Maiman, of Hughes Research Laboratories in Malibu, California, was the first to achieve success. He used a photographer's flash lamp to excite chromium atoms in a synthetic ruby crystal, which he selected because he had carefully observed how it absorbed and emitted light and determined that it could act like a laser. From a ruby rod, he produced red pulses the size of a fingertip on May 16, 1960. In December 1960, William Bennett, Jr., Ali Javan, and Donald Herriott at Bell Labs produced the prototype of a gas laser, which used a mixture of helium and neon to produce a continuous infrared beam. The first laser based on semiconductor means was created in 1962 by Robert N. Hall and coworkers at the General Electric Research and Development Center in Schenectady, New York.

Although lasers captured the public imagination quickly, perhaps due to their resemblance to science fiction's "heat rays," realistic implementations took years to create. While working on the ruby laser with Maiman, a young physicist named Irnee D'Haenens joked that the system was "a solution looking for a problem," a line that lingered in the laser community for many years. Townes and Schawlow anticipated that laser beams would be used in basic science and to transmit signals across the air or space. More strong beams, Gould imagined, would be capable of cutting and drilling a wide range of materials. Emmett Leith and Juris Upatnieks, two researchers at the University of Michigan, used lasers to create the first three-dimensional holograms in late 1963. The first lasers with widespread commercial applications were helium-neon lasers. They found immediate use as projecting straight lines for alignment, surveying, building, and irrigation since they could be modified to produce a visible red beam instead of an infrared beam. Eye surgeons soon used Ruby laser pulses to weld detached retinas back into place without cutting into the eye. The laser scanner for automated checkout in supermarkets, which was created in the mid-1970s and became popular a few years later, was the first large-scale application for lasers. Personal computers were followed by compact disc audio players and laser printers. Due to their wide range of applications, lasers have become popular. Laser pointers in lecture halls illuminate presentation points, and laser target designators direct smart bombs to their intended targets. Lasers can be used to solder razor blades, writing designs on production line items without touching them, remove excessive hair, and bleach tattoos. The asteroid Eros and Mars' surfaces were profiled in unparalleled detail by laser rangefinders in space probes. Lasers have assisted physicists in the laboratory in cooling atoms to within a fraction of a degree of absolute zero.

Fundamental Principles

Energy levels and **stimulated emissions**

The laws of quantum mechanics shape laser emission, restricting atoms and molecules to discrete amounts of stored energy that differ depending on the atom or molecule's nature. When an atom's electrons are all in the closest possible orbits to its nucleus, it has the lowest energy level. The ground state is the name given to this state. As one or more electrons in an atom absorb energy, they will travel to the outer orbits, and the atom is considered "excited." Excited states are rarely stable; when electrons fall from higher to lower energy levels, they emit extra energy in the form of light.

This emission could be generated in two ways, according to Einstein. Usually, isolated packets of light known as photons are released naturally and without external assistance. On the other hand, a passing photon might induce an atom or molecule to emit light if the energy of the passing photon exactly matched the energy that an electron would spontaneously release when falling to a lower-energy configuration. The ratio of lower-energy to higher-energy configurations determines which process takes precedence. Lower-energy configurations are more common than higher-energy configurations. This suggests that absorbing a photon and raising an electron from a lower-energy to a higher-energy configuration is more likely to occur than stimulating a higher-energy configuration to drop to a lower-energy configuration by releasing a second photon. Stimulated pollution will fade away as long as lower-energy states become more prevalent. If higher-energy configurations predominate (a phenomenon known as population inversion), randomly released photons are more likely to stimulate additional emissions, resulting in a photon cascade. Heat alone isn't enough to trigger a population inversion; something else has to excite the atoms or molecules selectively. This is usually accomplished by shining bright light on the laser material or sending an electric current through it. The most straightforward machine imaginable, such as Townes' ammonia maser, has just two energy levels. Three or four energy levels are more useful in laser systems. In a three-level laser, the material is first excited to a short-lived high-energy state. It then spontaneously drops to a slightly lower-energy state, known as a metastable state, with an extremely long lifetime.

Quantum Physics for Beginners

The metastable state is critical because it traps and retains excited energy, forming a population inversion that can be further stimulated to emit radiation, returning the species to its ground state. A three-level laser, such as Theodore Maiman's ruby laser, is an example.

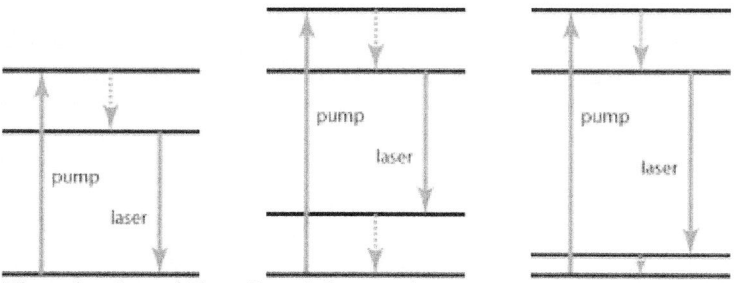

Three levels and Four Levels lasers picture

Regrettably, the three-level laser only operates if the ground state is zero. Most three-level lasers can only produce pulses because when atoms or molecules emit light, they concentrate in the ground state to absorb the stimulated emission and shut down laser activity.

The four-level laser came into the picture to solve this problem. It has an additional transition state between metastable and ground states. This enables several four-level lasers to emit a continuous beam for days.

Laser elements

It is possible to have Population inversions in a gas, liquid, or solid, but gases and solids are the most common laser media. Laser gases are usually found in cylindrical tubes and are excited by an electric current or an external light source, referred to as "pumping" the laser. Solid-state lasers, on the other hand, can rely on semiconductors or transparent crystals containing low concentrations of light-emitting atoms.

An optical resonator is needed to increase the light energy in the beam. The resonator is created by aligning a pair of mirrors so that light emitted along the line between them is reflected back and forth. Light reflected back and forth increases in intensity with each passing through the laser medium as a population inversion is formed in the medium. Other light bounces off the mirrors and is not intensified. Either or both mirrors relay a fraction of the incident light in a real laser cavity. The fraction of light transmitted, that is, the laser beam, depends on the type of laser. The sum of light added by stimulated emission on each round trip between the mirrors equals the light emerging in the beam plus losses within the optical resonator if the laser produces a continuous beam.

A laser oscillator is formed by combining a laser medium and a resonant cavity, which is commonly referred to as a laser. Many laser properties are determined by oscillation, which indicates that the system produces light internally. A laser will be nothing more than an optical amplifier without mirrors and a resonant cavity, amplifying the light from an external source but not producing a beam internally. In 1961, Elias Snitzer, an American

Optical researcher, demonstrated the first optical amplifier, but such devices were seldom used until the widespread acceptance of fiber optic communications.

Laser beam characteristics

Laser light differs from other light in that it is concentrated in a narrow beam, has a small spectrum of wavelengths (often referred to as "monochromatic"), and is made up of waves that are in phase with one another. The interactions between the stimulated emission phase, the resonant cavity, and the laser medium give rise to these properties. Stimulated emission releases a second photon of the same phase, wavelength, and direction as the one that stimulated the emission. The two are coherent with each other and also with peaks and valleys in phase. After that, both the original and new photons will induce the emission of more similar photons. The uniformity of the beam is improved by passing it back and forth through a resonant cavity, with the degree of coherence and beam narrowness depending on the laser configuration. Although a visible laser appears to create a point of light on the opposite wall of a room, the beam's alignment, or collimation, isn't perfect. The amount of beam radiating is determined by the distance between laser mirrors and diffraction and the scattering light at the aperture's edge. The laser wavelength divided by the dimension of the emitting aperture determines diffraction; the broader the aperture, the slower the beam spreads. A red helium-neon laser emits at a wavelength of 0.633 micrometers from a one-millimeter aperture, producing a beam that diverges at an angle of around 0.057 degrees, or one milliradian. At a distance of one kilometer, such a slight angle of divergence will create a one-meter spot. On the other hand, a standard flashlight beam creates a similar one-meter spot within a few meters. However, not all lasers emit narrow beams. Semiconductor lasers emit light, with a wavelength that nears one micrometer from an aperture of comparable duration, resulting in a divergence of 20 degrees or more. This necessitated the use of external optics to focus their beams.

The output wavelength is determined by the laser content, the stimulated emission process, and the laser resonator optics. A material may support stimulated emission over a limited range of wavelengths for each transition between energy levels; the width of that range varies on the nature of the material and the transition. The likelihood of stimulated emission depends on wavelength, and the process concentrates emission at wavelengths with the highest probability.

Resonant cavities enable laser oscillation at wavelengths that satisfy a condition of resonance: the number N of wavelengths must have the same value as the distance light travels between the mirrors on a round trip. The round-trip distance has to obey the formula $2L = N/n$ if L's cavity length and the material refractive index in the laser cavity are n. A longitudinal mode is a name given to each resonance. Even in semiconductor lasers, cavities are thousands of wavelengths long, so neighboring modes' wavelengths are close to each other. The laser typically emits light on two or more wavelengths that are within 0.1 percent of each other at the same time. For most practical applications, these beams are monochromatic; however, other optics may be applied to reach a single longitudinal mode laser oscillation and an even tighter range of wavelengths. The best laboratory lasers have a wavelength range that varies by less than 0.0000001%.

The smaller the wavelength spectrum, the more coherent the beam; that is, every light wave in the beam is in complete harmony with every other one. A quantity called coherence length is used to calculate this. Coherence lengths usually vary from millimeters to meters; for example, long coherence lengths are needed to record holograms of three-dimensional objects.

The most powerful experimental lasers can produce pulsed or continuous beams with average powers ranging from microwatts to over a million watts. The constant red beam from a laser pointer is an example of a continuous-wave laser; its output is nominally consistent for seconds or longer. Pulsed lasers channel their output energy into high-power bursts that last for just a short time. These lasers can shoot single pulses or a sequence of pulses at regular intervals. At the top of a very short pulse, instantaneous power may be very high. In laboratory experiments, pulses can be compressed to extremely short durations of around five femtoseconds (5×10^{-15} second) to "freeze" activity during events that occur quickly, such as stages in chemical reactions. Laser pulses may also be directed to focus high powers on small areas, similar to how a magnifier focuses sunlight into a small spot to ignite paper.

Types Of Lasers

Laser beams can be produced by crystals, glasses, semiconductors, gases, liquids, high-energy electron beams, and even gelatin doped with suitable materials. Hot gases near bright stars can emit strong stimulated emission at microwave frequencies in nature, but these gas clouds lack resonant cavities, so no beams are produced.

An external source of light excites atoms known as dopants applied to a host material at low concentrations in crystal and glass lasers, such as Maiman's first ruby laser. Glasses doped with erbium or ytterbium and glasses and crystals doped with the rare-earth element neodymium and can be drawn into fibers and used as fiber-optic lasers or amplifiers, which are good examples. Titanium atoms doped into synthetic sapphire can emit stimulated emission over a very large wavelength spectrum, which is why they're used in wavelength-tunable lasers. A variety of gases can be used as laser media. The standard helium-neon laser includes a small amount of neon and a lot of helium. Helium atoms absorb energy from moving electrons in the gas and transfer it to neon atoms, which emit light. The most well-known helium-neon lasers generate red light, but they can also produce yellow, orange, green, or infrared light, with standard powers in the milliwatt range. At visible and ultraviolet wavelengths, argon and krypton atoms stripped of one or two electrons will produce milliwatts to watts of laser light. The carbon-dioxide laser, which can produce kilowatts of continuous power, is the most effective commercial gas laser. Semiconductor diode lasers, which emit infrared or visible light when an electric current passes through them, are the most commonly used today. The emission occurs at the interface between two regions doped with different materials (p-n junction). If positioned within a suitable cavity, the p-n junction will serve as a laser medium, generating stimulated emission and providing lasing action.

Light oscillates in the junction plane in traditional edge-emitting semiconductor lasers that have mirrors on opposite sides of the p-n junction. Light resonates perpendicular to the p-n junction in vertical-cavity surface-emitting lasers (VCSELs), which have mirrors above and below the p-n junction. Its composition determines the wavelength of a semiconductor compound. In science, a few other forms of lasers are used.

The laser medium in dye lasers is a liquid containing organic dye molecules that can emit light at various wavelengths; changing the laser cavity adjusts the output wavelength or tunes it. Chemical lasers are gas lasers in which excited molecules emit stimulated emission due to a chemical reaction. Stimulated emission occurs in free-electron lasers as electrons travel through a magnetic field that regularly varies in direction and strength, causing the electrons to accelerate and emit light energy. Some researchers doubt whether a free-electron laser can be called a laser because the electrons do not transition between well-defined energy levels, but the term has stuck. Free-electron lasers can be tuned through a wide range of wavelengths depending on the energy of the electron beam and changes in the magnetic field. High-power lasers can be created by both free-electron and chemical lasers.

Laser Applications

Lasers produce light beams that are coherent, monochromatic, well-controlled, and precisely guided. While lasers are ineffective for general illumination, they are ideal for focusing light in space, time, or specific wavelengths. Many people were first exposed to lasers via laser light shows at concerts in the early 1970s. They were rotating laser beams of various colors that projected shifting patterns on planetarium domes, concert hall walls, or outdoor clouds.

The majority of laser applications fall into one of three categories: (1) information transmission and processing, (2) precise energy distribution, and (3) alignment, calculation, and imaging. These categories cover a wide range of uses, from precise energy delivery for delicate surgery to heavy-duty welding and from routine ceiling alignment to atomic property measurements in the lab.

Transmission and processing of information
Laser scanners

Lasers are valuable tools in telecommunications and information processing because they can concentrate laser beams on tiny spots and turn them on and off billions of times per second. A spinning mirror scans a red beam in laser supermarket scanners as clerks pass packages around the beam. Optical sensors detect light emitted from striped bar codes on products, decipher the symbol, and send the data to a device, which adds the bill's price. To play music, view video recordings, and read computer software. Small, low-cost semiconductor lasers read data from an increasing number of optical compact disc formats.

Around 1980, infrared lasers were used to create audio compact disks, and CD-ROMs (compact disc read-only memory) for computer data followed shortly after.
More efficient lasers are used in newer optical drives to record data on light-sensitive discs such as CD-R (recordable) or CD-RW (read/write), which can be played as common CD-ROM drives. DVDs (digital video or versatile disks) operate similarly. Still, they read more minor spots with a shorter-wavelength red laser, enabling the discs to carry enough information to play a motion picture. Blu-ray disks use blue-light lasers to read and store data at even higher densities.

Fiber-optic communication systems
Semiconductor laser beams are also used in fiber-optic communication networks to send signals over long distances. The optical signals are transmitted at infrared wavelengths of 1.3 to 1.6 micrometers, which are the most transparent wavelengths for glass fibers.
Optical fibers are the crucial elements in the global telecommunications network and most phone calls that fly outside the borders of a single town partly pass through them.

Precise delivery of energy

Industrial uses

Laser energy may be concentrated in space and time to melt, burn away, or vaporize a wide variety of materials. Even though a laser beam's overall energy is tiny, the focused power in small spots or at short intervals can be immense. While lasers are far more expensive than mechanical drills or blades, their unique properties allow them to accomplish tasks that would otherwise be impossible. Since a laser beam does not bend flexible materials like a mechanical drill, it can drill holes in soft rubber nipples for baby bottles. Laser beams can also drill or cut into incredibly tough materials without drilling bits or blades. Lasers, for example, have drilled holes in diamond dies that are used to draw wire.

Medical applications

Using a laser to cut tissue is a physical procedure similar to commercial laser drilling. Since their infrared rays are strongly absorbed by the water that makes up the bulk of living cells, carbon-dioxide lasers burn away tissue. The wounds are cauterized by a laser beam, preventing bleeding in blood-rich tissues, the female reproductive tract, and the gums. Laser wavelengths of less than one micrometer will reach the eye, reattaching a damaged retina or cutting internal membranes that sometimes become cloudy after cataract surgery. In diabetic patients, less-intense laser pulses may kill irregular blood vessels that extend across the retina, delaying the blindness that is often associated with the disease. Ophthalmologists correct vision problems by surgically removing tissue from the cornea and reshaping the eye's translucent outer layer with extreme ultraviolet pulses. Laser light can be transmitted to areas inside the body that the beams could not otherwise penetrate using optical fibers close to the tiny strands of glass that hold information in telephone systems. Threading a fiber through the urethra and into the kidney, for example, enables the end of the fiber to transmit strong laser pulses to kidney stones. The laser energy breaks up the stones into small enough pieces that can move into the urethra without the need for surgery. During arthroscopic surgery, fibers may also be implanted through small incisions to transmit laser energy to specific locations in the knee joint. The treatment of skin disorders is another medical use for lasers. Pulsed lasers can be used to disinfect tattoos and dark-red birthmarks known as port-wine stains. Unwanted body hair and wrinkles can be removed with cosmetic laser treatments.

High-energy lasers

Lasers can concentrate remarkably high powers in either pulses or continuous beams, according to scientists. Fusion science, nuclear weapons testing, and missile defense are all major applications for these high-power levels. To force atomic nuclei to fuse and release energy, extremely high temperatures and pressures are required. In the 1960s, physicists at California's Lawrence Livermore National Laboratory calculated that powerful laser pulses could create certain conditions by heating and compressing tiny pellets containing hydrogen isotope mixtures. They proposed using these "micro implosions" to produce energy for civilian use and model the implosion of a hydrogen bomb, which requires similar processes. Since then, Livermore has designed a series of lasers, specifically for the US government's nuclear weapons program, to test and refine these theories. Military laser weapon production began in the 1960s but gained little recognition until President Ronald Reagan initiated the Strategic Defense Initiative in 1983. High-energy lasers provide a means of transmitting destructive energy to targets at the speed of light, which is particularly appealing for fast-moving targets such as nuclear missiles. Military researchers have studied high-energy lasers for use as weapons on the ground, sea, air, and space, but none have been put in orbit. According to experiments, massive lasers can produce high forces; however, the atmosphere distorts such intense beams, causing them to be outspread and miss their targets. When the Cold War ended, these concerns hindered laser weapon research, though interest in laser weapons to protect against smaller-scale missile attacks persists.

Alignment, measurement, and imaging

Surveying

Construction workers and surveyors use laser beams to draw straight lines in the air. The beam is not visible in the air unless it is scattered by dust or haze, but it shines a bright point on a distant target. To determine the orientation and angle of the beam, surveyors bounce it off a mirror. A rotating beam can define a smooth plane for construction workers installing ceilings or walls, and it can set an angle for grading the irrigated property. Pulsed laser radar works in the same way as microwave radar in that it measures distance by timing a laser pulse takes to bounce back from a distant target. For example, laser radar accurately calculated the distance between the Earth and the Moon in 1969. Military laser range finders were developed in the 1970s to measure the distance between battlefield targets. For remote sensing, laser range finders are now commonly used. The Mars Global Surveyor used a laser altimeter to map elevations on the Martian surface. Aircraft-borne instruments can profile the layers of vegetation in a forest.

Interferometry and holography

Interferometry and holography require laser light coherence because they rely on interactions between light waves. In this way, they can make exact measurements and capture three-dimensional images. Their relative phases determine the outcome of combining light waves. If their peaks coincide with each other's valleys, they will interfere destructively, canceling each other out. If their peaks align, they will interfere constructively, producing a bright spot by splitting a beam into two identical halves that follow different directions. Changing one direction by half a wavelength would cause the two to be out of phase, resulting in a dark spot. This method has proven to be extremely useful for measuring tiny distances.

Holograms are created by splitting a laser beam into two halves and illuminating an object with one of them. In the plane of a photographic plate, this object beam is combined with the other half, the reference beam, to create a random-looking pattern of light and dark zones that record the light's wavefront from the object. As laser light illuminates the pattern from the same angle as the reference beam, it is dispersed, resulting in the reconstruction of a similar wavefront of light, which appears to the viewer as a three-dimensional representation of the target. Holograms can now be mass-produced using an embossing method similar to that used on credit cards, and they are no longer required to be displayed under laser light.

Research tool

The ability to precisely monitor laser wavelength and pulse duration has proven invaluable in fundamental physics and other sciences study. Lasers also played a crucial role in spectroscopy, the study of the light absorbed and released when atoms and molecules move between energy levels, which can expose atoms' inner workings. Lasers can focus a lot more power on a small range of wavelengths than other light sources, making them ideal for spectroscopic analysis.

The effects of spontaneous movements of atoms or molecules in a gas, for example, can be canceled by simultaneously illuminating samples with laser beams coming from opposite directions. This technique greatly enhanced the Rydberg constant measurement accuracy, which is crucial in atomic property calculations, and won Arthur Schawlow a share of the 1981 Nobel Prize in Physics. Other forms of high-precision laser spectroscopy were developed by Nicolaas Bloembergen, who shared the prize.

Laser spectroscopy has come a long way from those early days. On time scales faster than atomic motions in a molecule, laser pulses have been used to catch snapshots of chemical reactions as they occur. Ahmed Zewail won the Nobel Prize in Chemistry in 1999 for these methods, which have provided chemists new ways to understand chemical physics.

Physicists have also used laser beams' subtle forces to slow and capture atoms, molecules, and small particles. A Bell Labs researcher named Arthur Ashkin demonstrated that a tightly focused horizontal laser beam could trap atoms in a region with the highest light intensity, a technique known as "optical tweezers" used in various research.

Other research has shown that if the wavelength of a laser is calibrated to a point slightly off the wavelength of peak absorption, it can delay the motion of atoms. The atoms absorb photons from the beam repeatedly before releasing photons in random directions. The momentum of the photons delays the acceleration toward the laser beam. Placing the atoms at the intersection of six laser beams aimed at right angles slows their momentum in all directions, resulting in a clump of atoms less than 0.001 degrees above absolute zero. A magnetic field increases confinement and can lower temperature to less than one-millionth of a degree above absolute zero. These techniques contributed to developing a new state of matter known as a Bose-Einstein condensate. They awarded the 1997 Nobel Prize in Physics to Steven Chu, Claude Cohen-Tannoudji, and William D. Phillips.

CHAPTER 17:

LED

The term "light-emitting diode" refers to a device that emits light. A diode is an electrical component that allows only one direction of electric current to pass through it.

An LED is made up of two layers of crystal glued together tightly and processed in a controlled, clean environment to ensure that each layer has the exact chemical composition it requires.

LEDs emit light when the electrons that make up the electric current travel from one substance into the other when they are attached to a circuit the right way up (so that the electric current can pass through).

Physicists have used quantum mechanics principles to explain better how electrons behave when they move through crystalline materials. This has helped them to create LEDs that emit a variety of colors of light.

Traditional LEDs have been available for decades and emit a single color, such as red or green. On the other hand, blue LEDs proved extremely difficult to make due to the difficulty of obtaining the required ingredients. Three Japanese researchers shared the Nobel Prize in Physics in 2014 for their breakthroughs in solving the problems that have stopped blue LEDs from being a possibility.

In most cases, white light is desired in a structure so that a lamp may carry three different LEDs, such as green, red, and blue. White light is generated by combining these colors. You may also create a white LED by exciting a chemical layer inside the bulb with blue light, allowing it to emit white light. In houses, each technique may create adequate LED lighting that is much more efficient than other forms of lighting.

This is particularly significant for the 1.5 billion people who do not have access to electricity grids because solar panels could power the lights. Lighting is projected to consume 20% of global electricity; however, if all lights were LED, this figure would drop to 4%.

Working under bright, white lighting all day can be dangerous, so some LED lighting systems are built to produce bluer light in the morning and redder light later in the day. This causes less disruption to the body's natural everyday rhythm.

LED lighting's advantages:

- LED lamps are produced with a solid material that is more durable than filament or fluorescent lights and lasts much longer.
- Costs and carbon emissions are decreased over time.
- For LiFi wireless computer networks, it is possible to use LED lights.

CHAPTER 18:

QUANTUM COMPUTING

Quantum computing will change the world as it can revolutionize medicine, decryption, communications, and artificial intelligence. Corporations are producing quantum computers. Also, Countries have put billions of dollars into the project. A rush to quantum superiority has started. A quantum machine has already outperformed a conventional one. But, first and foremost, what is quantum computing? What's more, how does it work?

What is quantum computing?

Let's start with the fundamentals.
Bits are used in a standard computer chip. These are tiny switches that can be either off (represented by a zero) or on (represented by a one) position. Every app you use, every website you visit, and every photograph you take comprises millions of these bits in some one-to-one correspondence. This is fine for most things, but it does not represent how the world functions. Things aren't necessarily on or off in nature. They are unsure. Also, our most powerful supercomputers struggle to cope with ambiguity. That's a problem. That's because physicists have learned over the last century that when you get down to a microscopic scale, strange things start to happen. To explain them, they've created an entirely new field of research. It's referred to as quantum mechanics. Quantum mechanics underpins physics, which in turn underpins chemistry, which in turn underpins biology. Scientists would need a more straightforward way to make calculations that can accommodate uncertainty to model all of these things accurately. Quantum computers are here to help.

How do quantum computers work?

Quantum computers use qubits instead of bits. Qubits may be in what's known as 'superposition,' where they're both on and off at the same time, or anywhere on a continuum between the two, rather than either being on or off.
Take a coin and toss it. It can be either tails or heads if you turn it. However, if you spin it, it has a chance of falling on heads or tails. It can be either until you weigh it by stopping the coin. One of the aspects that makes quantum computers so strong is superposition, similar to a spinning coin. Uncertainty is permitted within a qubit.
If you ask a standard machine to find its way out of a maze, it will try each branch one at a time, ruling them out one by one before finding the correct one. A quantum computer will navigate all of the maze's paths at the same time. It can keep doubt in its mind. It's like sifting through the pages of your best adventure novel. Instead of restarting the book, if your character dies, you can take a new direction right away. Entanglement is something else that qubits can do. Usually, when two coins are flipped, the outcome of one does not

affect the result of the other. They're self-sufficient. Even if two particles are physically apart, they are bound together in entanglement. If one of them comes up as heads, the other would as well.

It sounds magical, and scientists are still confused as to how or why it works. However, in the sense of quantum computing, it means that knowledge can be passed around even though it involves uncertainty. You can use the spinning coin to perform complicated calculations. And if you can connect several qubits, you can solve problems that would take millions of years for our best computers to solve.

What can quantum computers do?

Quantum computers are about more than just getting things done quicker and more reliably. They'll allow us to do things we couldn't have imagined without them. Stuff that not even the most efficient supercomputer can do.

They can propel artificial intelligence forward at a breakneck speed. Big Tech Corporations are also using them to develop self-driving car applications. They'll be necessary for simulating chemical reactions as well. Supercomputers can currently only study the simplest molecules. However, quantum computers use the same quantum properties as the molecules they're simulating. They should be able to manage even the most complex reactions with ease. This may mean more effective goods, such as new battery materials for electric vehicles, safer and less costly medications, and significantly improved solar panels. According to scientists, quantum simulations can also aid in the discovery of a cure for Alzheimer's disease.

Quantum computers can be helpful when a big, unknown, and complex system needs to be simulated. Quantum computing can be used to explain quantum physics in several ways, including forecasting financial markets, enhancing weather forecasts, and modeling the behavior of individual electrons.

Another important application would be cryptography. Many encryption schemes currently depend on the complexity of breaking large numbers down into prime numbers. This is known as factoring, and it is a long, costly, and inefficient process for traditional computers. Quantum computers, on the other hand, can do it quickly. And this could jeopardize our data.

There are rumors that intelligence agencies worldwide are now stockpiling large volumes of encrypted data assuming that they will crack it with a quantum computer one day. Quantum encryption is the only way to strike back. This is based on the uncertainty principle, which states that it is impossible to calculate anything without affecting the outcome. The keys to quantum encryption couldn't be copied or hacked. They'd be almost impenetrable.

When will I get a quantum computer?

A quantum chip would almost certainly never be found in your laptop or smartphone. The iPhone Q will not be released. Quantum computers have been theorized for decades, but their development has been slowed by being extremely sensitive to interference.

A qubit can be kicked out of its fragile state of superposition by almost everything. As a result, quantum computers must be kept free of all electrical interference and kept at temperatures close to absolute zero. That's colder than the farthest reaches of the universe. Academics and companies will primarily use them, and they will be accessed remotely. Corporations' quantum computer can be used via the company's website, and you can even play a card game with it.

However, a lot of effort will be put into quantum computers before they can do what they promise. The best quantum computers currently have about 50 qubits. That alone makes them highly efficient since every qubit added increases processing capacity exponentially. However, because of the intrusion issues, they have incredibly high error rates.

They're strong, but they're also unreliable. For the time being, statements of quantum superiority must be treated as a grain of salt. One big corporation released a paper in October 2019 claiming to have reached quantum dominance – the stage at which a quantum machine would outperform a classical computer. However, competitors challenged the argument, stating that they had not even thoroughly utilized the power of modern supercomputers. So far, most breakthroughs have occurred in managed environments or with problems for which we already have a response. In any case, gaining quantum dominance does not suggest that quantum computers can perform practical tasks. Researchers have made significant progress in designing quantum computer algorithms. However, the devices themselves still need a great deal of improvement.

CHAPTER 19:

SUPERCONDUCTIVITY

Heike Kamerlingh Onnes discovered superconductivity in 1911 while researching the properties of metals at low temperatures. He was the first to liquefy helium, which has a boiling point of 4.2 K (K is Kelvin Degrees) at atmospheric pressure, a few years before, and this had opened up a new range of temperature for experimental research. He was astounded to discover that the resistance of a small tube filled with mercury decreased from 0.1 at 4.3 K to less than $3*10^{-6}$ at 4.1 K when he tested its resistance. Mercury is a superconductor below 4.1 K, and no experiment has found any resistance to steady current flow in a superconducting material. The critical temperature T_c is the temperature below which mercury becomes superconducting. In 1913, Kamerlingh Onnes was awarded the Nobel Prize in Physics for his work on the properties of matter at low temperatures, which contributed to the production of liquid helium, among other things (Nobel Prize citation). Many more elements were found to be superconductors since this initial discovery. As the Periodic Table shows, superconductivity is far from a rare phenomenon. At the bottoms of the cells, the number shows the critical temperatures of the elements that become superconducting at atmospheric pressure, ranging from 9.3 K for niobium (Nb, Z = 41) to $3*10^{-4}$ K for rhodium (Rh, Z = 45). Orange cells are components that only become superconductors when exposed to intense pressures. Chromium (Cr, Z = 24) as thin films, Carbon (C, Z = 6) in the form of nanotubes, platinum (Pt, with Z = 78) as a compacted powder, and palladium (Pd, Z = 46) after irradiation with alpha particles, are the four pale pink cells that are superconducting in specific examples. Also, at the lowest temperatures possible, copper (Cu, Z = 29), silver (Ag, Z = 47), and gold (Au, Z = 79), three elements that, at room temperature, are excellent conductors, do not become superconductors.

Walter Meissner and Robert Ochsenfeld, in 1933, found out that superconductors are more than ideal conductors of electricity. They made a considerable contribution to our understanding of superconductivity. They also made it known that superconductors have the essential property of not allowing a magnetic field to move into their interior. The field is only removed if it is below a specific critical field power, depending on the sample, temperature, and specimen geometry. Superconductivity vanishes beyond this critical field power. In 1935, brothers Fritz and Heinz London suggested a model that explains the field's exclusion, but it took another 20 years for a microscopic theory to emerge. Three US physicists, Leon Cooper, John Bardeen, and John Schrieffer, published the long-awaited quantum theory of superconductivity in 1957. They received, in 1972, the Nobel Prize in Physics (it was called the BCS theory). According to their theory, the vibrations of the ion lattice mediate an enticing interaction between electrons in the superconducting state. Owing to this interaction, electron pairs become entangled, and all of the pairs condense into a macroscopic quantum state known as the condensate, which remains in the superconductor. In a superconductor, not all free electrons are in the condensate; those are known as superconducting electrons, whereas the rest are known as regular electrons. There are very few normal electrons at temperatures well below the critical temperature. Still, as the temperature rises, the proportion of regular electrons increases until all of the electrons are normal at the critical temperature. Since superconducting electrons are connected in a macroscopic state, they behave coherently. As a result, the coherence length (the Greek lower-case xi, pronounced "ksye") is a characteristic distance over which their number density will change.

Since scattering an electron from the condensate involves a large amount of energy – more than the available thermal energy to an electron below the critical temperature – superconducting electrons will flow without being scattered, i.e., without any resistance. The BCS theory successfully explained many known properties of superconductors, but it predicted an upper limit of roughly 30 K for the critical temperature.

In 1957, another significant theoretical advance was made. Alexei Abrikosov predicted the second form of superconductor, which would behave differently than lead and tin. When the applied field strength was poor, this new form of superconductor would expel the field from its interior. Still, over a broad range of applied field strengths, the superconductor would be threaded by normal metal regions through which the magnetic field could move. Because of the field's penetration, superconductivity could occur in magnetic fields of up to 10 T (Tesla) or more, allowing for a wide range of applications. Abrikosov, in 2003, was awarded the Nobel Prize in Physics for "pioneering contributions to the theory of superconductors and superfluids" for this work and subsequent research (Nobel Prize citation).

The early 1960s had made significant developments in superconductor technology. There was the discovery of alloys superconducting at temperatures greater than the critical temperatures of elemental superconductors. In particular, alloys of niobium and titanium (NbTi, T_c = 9.8 K) and niobium and tin (Nb 3 Sn, T_c = 18.1 K) were being commonly used to manufacture high-field magnets. This was because of the need for powerful magnets for particle accelerators. Around the same time, Brian Josephson made a key theoretical observation that had far-reaching implications for applying superconductivity on a microscale. A current could flow between two superconductors separated by a very thin insulating layer, he predicted. The so-called Josephson tunneling effect has been commonly used for sensitive measurements, such as determining fundamental physical constants and measuring magnetic fields a billion times weaker than the Earth's field. He received the Physics Nobel Prize for his theoretical predictions of the properties of a super current through a tunnel barrier; those phenomena are primarily known as the Josephson effects (Nobel Prize citation) in 1973.

Despite the BCS theory's prediction that the upper limit for T_c was less than 30 K, the search for superconductors with high critical temperatures continued decades after its publication. For scientists working in this area, the holy grail was a substance that was superconducting at liquid nitrogen temperature (77 K) or, even better, at room temperature. This would remove much of the technology and costs associated with using liquid helium for cooling, and superconductivity applications would become much more commercially feasible almost immediately. Georg Bednorz and Alex Muller made the breakthrough in 1986 when they discovered that ceramics made of barium, lanthanum, copper, and oxygen were superconducting at 30 K, the highest known critical temperature at the time. Since this material is an insulator at room temperature, the finding was wildly unexpected. The Nobel Prize for Physics was awarded to them the following year. The prize was awarded soon after their results that demonstrated the importance of their work was published.

As a result of this discovery, a research frenzy exploded, and a few other scientists started to study similar materials. Paul Chu discovered that replacing lanthanum with yttrium created a new ceramic material with a critical temperature of 90K in 1987. With the prospect of commercial feasibility for the latest materials, a rush to develop new high-temperature superconductors and justify why they super conduct at such high temperatures resulted. The highest critical temperature for a thallium-doped mercuric cuprate, $Hg_{0.8} Tl_{0.2} Ba_2 Ca_2 Cu_3 O_{8.33}$, was 138 K at the time of writing (2005). While no materials with significantly higher critical temperatures have been discovered in recent years, other important discoveries have been made. Among them is the discovery that ferromagnetism and superconductivity can coexist in various materials. Also, the first high-temperature superconductors without copper have been discovered. Scientists must constantly re-examine long-held hypotheses on superconductivity and accept new combinations of elements resulting from startling findings like these.

Unfortunately, no superconductors with critical temperatures above room temperature have yet been discovered, so cryogenic cooling remains a critical component of any superconducting application. Difficulties have hampered the production of new applications of high-temperature superconductors in fabricating ceramic materials into conducting wires or strips. Despite these disadvantages, industrial usage of superconductors is growing.

CHAPTER 20:

SUPERFLUIDS

Along time has passed since the groundbreaking physics discovery of superfluidity: more than 80 years. The capacity of a liquid to flow through narrow channels without visible friction is the most central concept of superfluidity. However, this is just one of a variety of intriguing characteristics. If we put a liquid in a bucket and slowly rotate it while cooling it into the superfluid phase, the liquid, which spins with the bucket at first, will appear to come to a halt. The Hess-Fairbank effect is the name given to this phenomenon. Superfluidity can now be explicitly observed in helium isotopes and ultra-cold atomic gases. Excitons, bound electron-hole pairs that can be found in semiconductors, exist in extraterrestrial systems like neutron stars. There is circumstantial evidence for their presence in other terrestrial systems like excitons.

Helium-4 was first liquefied in 1908, but it wasn't until 1936 and 1937 that scientists discovered it had properties unlike any other material known at the time, below the temperature of 2.17 degrees absolute now called the lambda point. The low-temperature process's thermal conductivity, now called He-II, is exceptionally high, implying a convection mechanism with unusually low viscosity. Pyotr Kapitza in Moscow and Don Misener and John Allen at the University of Cambridge conducted simultaneous measurements of the viscosity of helium found in a thin tube as a function of temperature in 1938. Both groups discovered a decrease in He-II that occurred intermittently at the lambda stage. Kapitza invented the term superfluidity for this behavior based on the analogy with superconductivity. Particles of the same kind are indistinguishable in the quantum universe, and there are only two types of particles: fermions and bosons. On the other hand, a composite boson is made up of an even number of interacting fermions. For example, an atom of helium-4 is a composite boson made up of six fermions (two protons, two neutrons, and two electrons). Helium-4 atoms undergo Bose-Einstein condensation and become superfluid at sufficiently low temperatures. In the BCS theory of superconductivity, electrons with an attractive interaction may combine to form charged composite bosons known as Cooper pairs, which condense to form a superconductor.

Superconductivity, in our current interpretation, is little more than just superfluidity in an electrically charged device. After commencing in a superconducting ring, a current can flow forever down a narrow capillary without apparent friction, just like how a superfluid liquid can flow forever down a thin capillary without apparent friction – or at least for a period much longer than the age of the Universe! The analog of the previously described Hess-Fairbank effect is a little less intuitive. When applying a magnetic field to a metal's surface, the natural, non-superconducting state has minimal effect. When a metal is superconducting, it causes an electric current, which is known as diamagnetism. In a thin ring, that would be the end of the story. Still, in a bulk sample, this current generates its magnetic field in the opposite direction of the external one, and the latter is gradually screened out of the metal entirely. This is known as the Meissner effect, and it creates fantastic effects like superconducting levitation.

Superfluid helium has only a few direct applications. The superfluid phase of helium-4 is an outstanding coolant for high-field magnets due to its exceptionally high thermal conductivity, and both isotopes have some uses, such as exotic particle detectors. Although superfluidity has other particular indirect benefits, it is most helpful in developing theory and understanding high-temperature superconductivity.

CHAPTER 21:

QUANTUM DOTS

What on earth is a quantum dot?

A quantum dot is a microscopic semiconductor particle (1–20 nanometers in size) with different properties than its larger counterpart due to its size. The properties change in terms of optical and electronic properties, and these properties can be altered on a given spectrum by changing the surface to volume ratio. This one-of-a-kind behavior in response to quantum confinement is what draws researchers' attention and gives them their names. When we give energy to these semiconductors or "excite," them the electrons inside pass, thereby creating a hole where the electron would usually be located, forming an electron-hole pair.

The electron returns to its original location when it loses energy. This energy that was previously lost by the electron is restored in the form of a photon of light. Manipulation of the given and lost energy results in emissions of different wavelength and frequency photons. These photons are visually displayed as different colors of light. Larger Quantum dots emit longer wavelength photons that appear as orange or red light. In contrast, smaller Quantum dots emit shorter-wavelength photons that appear as a blue or green light, allowing for a broad spectrum of colors depending on the manipulation. The properties of Quantum Dots were discovered in 1930, but it wasn't until the 1960s that researchers started to focus on light emissions and potential applications for this unusual behavior. Lighting, high-definition LCD televisions, clean energy, and medical diagnostics are only a few of the industries that use QDs today.

The Application in Medical Diagnostics

Research into the application of QDs in biotech was not pursued right away because of the toxicity of the cells. Zinc sulfide, cadmium selenide, lead sulfide, and indium phosphide are the most widely used components. Research into their use in medicine was initially dismissed because of the heavy metal content of these products and the risk of them leaking into the body, and the inability to combine them with biologically compatible solutions. However, in 1998, a water-soluble coating for the Quantum Dot was perfected, paving the way forward. This paved the way for further studies, and nearly 100,000 manuscripts about QD applications were written between the year 2000 and 2010, with 10% of them focusing on biotech applications. A QD can be encased in a coating that looks like organic receptors and then binds to disease biomarkers. This is visible in vivo due to the emitted light.

So, where will this be used, and what are the advantages over conventional methods?

QDs may be used to replace organic dyes that are currently used in in vivo diagnostics. These dyes, which bind to antibodies formed by the immune system in response to antigens, are organic by definition, which means they will break down and deteriorate over time.

They also have a narrow emission spectrum, which means there is a wider variety of subjective errors in interpreting data, and they propagate rapidly across human tissue. When a surgical tumor is removed, a dye is pumped into the infected area. So, a snapshot of the area is taken for the surgeon to operate. Since the dye bleeds, a more significant margin of error is used, ensuring that more healthy tissue can be moderately removed to guarantee that the tumor is completely removed.

When compared to QDs, it appears that current dyes fall short. The fluorescent yield of the QD is brighter. It has a smaller color range, which means colors differ on a broader scale and are easier to discern from one another. This is primarily due to the QD's ability to consume more energy than organic dyes. Since QDs are inorganic, they do not break down in the same way as dyes do, and it is possible to use one type of excitation source to generate results from multiple QDs. In contrast, dyes can only be activated by one kind of energy. QDs often pass more slowly through healthy tissue, and by using infrared light, the photon released is visible through it. As a result, less unaffected tissue and muscle mass must be extracted since the tumor's boundaries are more clearly defined.

Quantum dots can be used in two ways: directly (active) or indirectly (passive). The QDs are manufactured as detection molecules, and when applied to the sample, they bind to the target antigens, washing away those that do not bind, and the emitted light can then see the biomarker mass/cells. Since QDs react to their unique properties rather than the energy source, it is possible to build a massively multiplexed assay to detect a wide range of biomarkers in a single test. Multiplexed Optical Coding is the term for this process. Instead of emitting the energy lost as light, the indirect approach moves the energy lost from the QD to other nearby fluorescent molecules. The procedure is repeated until the energy lost is emitted as light from the secondary molecules rather than the QD. Forster Resonance Energy Transfer is the term for this (FRET).

The application of QDs to medical diagnostics tends to provide several advantages over current in vivo imaging options. Since QDs can bind to antigens/biomarkers, they're suitable for targeted drug delivery since the drug is kept between the inorganic center and the polymer coating. This is especially effective in circumstances where the dosage is harmful to healthy cells, such as chemotherapy.

They've also been shown to have antibacterial properties, as their presence disrupts the antioxidative system in cells, which can be used against bacteria in a dose-dependent manner.

Problems with Quantum Dots

The implementation of QDs, like any modern technology, is not without issues. Although the toxicity risk has been eliminated using a water-soluble coating, there are still concerns about the method's application – mainly for in vivo applications. The capacity of the kidneys to extract QDs from the body is a research subject since they are more significant than dyes (6–60nm vs. 0.5nm). Also, processing QDs in large amounts without variability and marking them is difficult. A method to do so is necessary to determine which antibodies are present, but there is no standard method.

There are some cases where the solution isn't always optimal in terms of drug delivery capabilities. Larger QDs have trouble entering solid tissues, and smaller QDs are being removed from the body by the kidneys before the dose can take effect in some cases. However, research into altering the status quo to raise the positives while eliminating the negatives continues.

Many businesses are involved in QDs and the use of nanotechnology in medical diagnostics. Quantum dots and their unique properties will continue to be researched by well-known diagnostic companies. Significant manufacturers and a growing number of nanotechnology firms are also looking into their application.

CHAPTER 22:

MRI

Magnetic Resonance Imaging (MRI) can be defined as a non-invasive imaging technique that provides accurate three-dimensional anatomical images. It's often used to identify diseases, diagnose them, and monitor their development. It uses cutting-edge technology to stimulate and track changes in the rotational axis of protons in the water that makes up living tissues.

Magnetic resonance imaging (MRI) makes detailed images of any part of the body by utilizing the body's magnetic properties. The hydrogen nucleus (a single proton) is employed for imaging since it is abundant in water and fat.

To make things easier, let's compare the hydrogen proton to the earth, which spins on its axis and has a north-south pole. It's similar to a little bar magnet in this regard. These hydrogen proton "bar magnets" rotate in the body with their axes randomly aligned.

The protons ' axes line up when the body is placed in a powerful magnetic field, such as an MRI scanner. The MRI scanner's axis is aligned with the magnetic vector formed by this consistent alignment. MRI scanners come in a range of field strengths, ranging from 0.5 tesla to 1.5 tesla.

The magnetic vector is deflected when extra energy (in a radio wave) is introduced to the magnetic field. The hydrogen (in this case) and the strength of the magnetic field determine the radio wave frequency (RF) that induces resonance into the hydrogen nuclei.

The strength of the magnetic field may be altered electronically from head to toe using a series of gradient electric coils, and by adjusting the local magnetic field in small increments, different slices of the body will resonate as different frequencies are applied.

The magnetic vector goes back to its resting state when the radiofrequency source is turned off, causing a signal (also a radio wave) to be emitted. This signal is used to generate magnetic resonance pictures. Receiver coils are placed around the bodily component to act as aerials, allowing the transmitted signal to be detected more easily. The received signal's intensity is then plotted on a greyscale, and cross-sectional images are created.

Multiple transmitted radiofrequency pulses might be employed in sequence to emphasize specific tissues or anomalies. When the transmitted radiofrequency pulse is turned off, various tissues relax at different speeds, resulting in a distinct emphasis. There are two approaches to calculate the time for the protons to relax fully. The first is the time the magnetic vector needs to return to rest, and the second is the time it takes for the axial spin to come to rest. T1 relaxation is the first, and T2 relaxation is the second.

Different tissues (such as fat and water) relax at different rates and are distinguished. As a result, an MR examination consists of a succession of pulse sequences. For example, utilizing a "fat suppression" pulse sequence will remove the signal from fat, leaving only the signal from any anomalies inside it.

Because a rise in water content characterizes most diseases, MRI is a sensitive diagnostic for disease detection. The specific nature of the pathology can be difficult to determine; for example, an infection and a tumor can appear similar in some circumstances. A radiologist's meticulous examination of the photographs will typically provide the proper answer.

MRI has no known biological risks since, unlike x-rays and computed tomography, it employs radiofrequency energy that is found all around us and does not harm tissue as it passes through.

Because of the potential for movement within a magnetic field, pacemakers, metal clips, and metal valves can be problematic in MRI scanners. Metal joint prostheses are less challenging, albeit there may be some picture distortion near the metal. MRI departments routinely inspect for implanted metal and provide advice on its safety.

CHAPTER 23:

BONUS CHAPTER: RELATIVITY

Albert Einstein's theory of relativity is famous for predicting some strange but accurate events, such as astronauts aging slower than Earthlings and solid objects shifting shape at high speeds.

But, if you pick up a copy of Einstein's original article on relativity from 1905, you'll find it a simple read. His writing is simple, and his equations are algebra—nothing that an average high school student would find difficult.

That's because Einstein was never interested in complicated math. He preferred to think visually, creating experiments in his thoughts and working them through until he could clearly understand the ideas and physical principles.

Here's how Einstein began his thought experiments when he was just 16 years old and how they led to the most revolutionary equation in contemporary physics.

Running alongside a Light Beam, 1895

By this time, Einstein's open contempt for his own Germany's rigorous, authoritarian educational systems had already gotten him tossed out of high school, so he traveled to Zurich in the hopes of enrolling at the Swiss Federal Institute of Technology (ETH).

First, Einstein decided to spend a year studying at a school in the nearby town of Aarau that emphasized avant-garde approaches such as independent thought and concept visualization. He soon found himself thinking about what it would be like to sprint beside a light beam in that cheerful environment.

Einstein had previously learned what a light beam was in physics class: a series of oscillating electric and magnetic fields rippling along at the observed speed of light, 186,000 miles per second. Einstein reasoned that if he ran beside it at such speed, he should be able to observe a set of oscillating electric and magnetic fields hanging right by him, seemingly motionless in space.

That, however, was impossible. For starters, such stationary fields would defy Maxwell's equations, which codified all scientists knew about electricity, magnetism, and light at the time. The rules were (and still are) very strict: any ripples in the fields had to travel at the speed of light and could not stop—no exceptions.

Worse, static fields would contradict the idea of relativity, which physicists have believed in since Galileo and Newton's time in the 17th century. In essence, relativity stated that the laws of physics could not be affected by your speed; all you could measure was the velocity of one item relative to another.

However, when Einstein applied this idea to his thought experiment, he discovered a contradiction: Relativity demanded that whatever he could observe while sprinting alongside a light beam, including static fields, be something Earthbound physicists could generate in the lab. But no one has ever seen anything like that before.

This difficulty would plague Einstein for another ten years through his university studies at ETH and his migration to Bern, Switzerland's capital city, where he worked as a patent examiner. He vowed to overcome the contradiction once and for all at that point.

Measurement of Light from a Moving Train, 1904.
It wasn't simple. Einstein tried every possible solution, but none of them worked. He began to investigate a basic but bold idea almost out of desperation. He reasoned that while Maxwell's equations might work for everyone, the speed of light remained constant.

In other words, it didn't matter if the light beam's source was moving toward you, away from you, or off to the side when you saw it zip by, nor how fast it was moving. The velocity of such a beam would always be 186,000 miles per second. Einstein would never notice the fixed, oscillating fields because he could never catch the light beam, among other factors.

Einstein saw no other way to reconcile Maxwell's equations with the concept of relativity than this. However, this remedy appeared to have its fatal defect at first. Imagine shooting a light beam along a railroad embankment at 2,000 miles per second while a train roars by in the opposite direction.

Standing atop the barrier, someone would estimate the light beam's speed to be 186,000 miles per second. However, someone on the train would notice it passing by at a speed of only 184,000 miles per second. Einstein argued that the concept of relativity would be violated if the speed of light was not constant because Maxwell's equations would have to look different inside the railway carriage.

For nearly a year, Einstein was perplexed by this seeming contradiction. He was walking to work with his best friend Michele Besso, an engineer he had known since their college days in Zurich, on a sunny morning in May 1905. As they often did, the two men were discussing Einstein's predicament. Suddenly, Einstein recognized the solution. "Thank you," Einstein told Besso when they met the following day after he had worked on it overnight. "I've fixed the problem totally."

Lightning Strikes a Moving Train in May 1905
According to Einstein, observers in relative motion experience time differently: it is entirely feasible for two events to occur simultaneously from one observer's perspective but different times from the others. And both observers are correct.

Another thought experiment by Einstein later demonstrated this idea. Assume you're standing on a railway embankment, watching a train rumble by. This time, though, a bolt of lightning strikes each end of the train as it passes through the halfway. Because the lightning strikes are at the same distance from the observer, their light arrives simultaneously in his eye. So, he's right when he says they happened at the same time.

Meanwhile, another train passenger is sitting precisely in the middle of the train. From her perspective, the light from both strikes must go the same distances, and she will measure the speed of light in both directions to be the same. However, because the train is moving, the light from the rear-facing lightning must travel further to catch up, so it arrives a few seconds later than the light from the front. She can only conclude that the strikes were not simultaneous because the light pulses arrived at various times.

In other words, Einstein discovered that simultaneity is relative. The odd effects we now associate with relativity are a straightforward algebra problem once you accept that.

Einstein rushed through his ideas and submitted his manuscript for publication only a few weeks later. "On the Electrodynamics of Moving Bodies," he titled it, alluding to his struggle to reconcile Maxwell's equations with the theory of relativity. And he wrapped it up with a thank you note to Besso ("I am thankful to him for several valuable recommendations"), ensuring his friend's longevity. Mass and Energy, September 1905

However, that initial paper was only the beginning. Throughout the summer of 1905, Einstein was obsessed with relativity, and as an afterthought, he submitted a second paper in September.

It was the result of yet another mental exercise. Consider a stationary item, Einstein explained. Now imagine it emits two similar light pulses in opposite directions. The object will remain in place, but its energy content will diminish because each pulse releases energy. What would this process appear like to a moving spectator, Einstein wondered? The object would keep going in a straight path from her perspective while the two pulses flew away. Even though the two pulses' speeds are the same—the speed of light—their energy is not: The pulse that moves ahead in the direction of motion now has more energy than the one that moves backward.

With a little more algebra, Einstein demonstrated that for all of this to be consistent, the object had to lose not only energy but also mass when the light pulses faded. Alternatively, mass and energy are interchangeable terms.

Einstein devised a formula that connects the two. He created the most famous equation ever written using today's notation, abbreviating the speed of light with the letter c: $E = mc^2$.

CONCLUSIONS

Congratulations on having completed this journey. Quantum Physics is such a fascinating topic. In this book, we tried to give just an introduction. We went through its origin, and we arrived to talk about the daily applications, tools that affect our everyday life. There are so many other experiments, studies, researches that are ongoing and could lead to discoveries that will enhance human knowledge to the next levels. Suppose you are interested in the topic and the math and models under it. Then, in that case, I suggest you start studying academic books and eventually go to University and College. Thank you for choosing this book, and keep learning. If you enjoyed it, please do not forget to leave a review on the store.

Live long and prosper!

Printed in Great Britain
by Amazon